Synthesis Lectures on Mathematics & Statistics

Series Editor

Steven G. Krantz, Department of Mathematics, Washington University, Saint Louis, USA

This series includes titles in applied mathematics and statistics for cross-disciplinary STEM professionals, educators, researchers, and students. The series focuses on new and traditional techniques to develop mathematical knowledge and skills, an understanding of core mathematical reasoning, and the ability to utilize data in specific applications.

Dušan Stipanović

Difference Equations and Machine Learning

Springer

Dušan Stipanović
Department of Industrial and Enterprise
Systems Engineering and Coordinated Science
Laboratory
University of Illinois Urbana-Champaign
Urbana, IL, USA

ISSN 1938-1743 ISSN 1938-1751 (electronic)
Synthesis Lectures on Mathematics & Statistics
ISBN 978-3-032-00909-8 ISBN 978-3-032-00910-4 (eBook)
https://doi.org/10.1007/978-3-032-00910-4

This Springer imprint is published by the registered company Springer Nature Switzerland AG
The registered company address is: Gewerbestrasse 11, 6330 Cham, Switzerland

Dedicated to Mama Ljilja, for unconditional love, support, and devotion.

Preface

There has been a long inner struggle of mine to write a monograph. Then, came the opportunity to write a monograph on a topic of my choice. For largely unexplainable reasons, I chose and proposed to write a monograph on difference equations and machine learning. Essentially, this monograph does contain technical material on stability and control of difference equations with a particular application to recurrent neural networks and machine learning. The particular application is believed to be related to a regulation type of control that was inspired by a modified internal model control approach. Not more than that but certainly less. This monograph is a result of an interplay among beliefs, struggles, experiences, expertise, doubts, excitements, imaginations, preconditioning, and insecurities.

Beograd, Serbia/Kvarner, Croatia/Urbana,
USA/Walnut Creek, USA
June 2025

Dušan Stipanović

Acknowledgments This book would not "see the light of day" without my compatriot from "the balcony" Prof. Boris Murmann. Boris initially introduced me and then encouraged me to start investigating dynamic behaviors of long short-term memory neural networks, producing very interesting dynamical behaviors and very popular due to their successful applications in machine learning.

My work on recurrent neural networks as well as my research in general has been greatly supported by Prof. Claire Jennifer Tomlin. Claire has streamlined my ambiguities and with great patience helped me overcome my professional insecurities. She genuinely supported my interests and endeavors regardless of their plausibility.

I was introduced to neural networks and internal model control regulation by my undergraduate advisor, late Prof. Miroslav Mataušek. I worked with him on my undergraduate thesis while the war in former Yugoslavia was in full swing. Working with him in those uncertain days, was not only inspiring but also kept my thoughts away from tragic events that also plagued all the innocent people living in my home country of Yugoslavia. Sadly valuable, that revealed to me some of our darkest natures and working on an interesting technical problem was not only engaging but also an escape from the harsh reality. My journey into learning and controlling complex systems continued with Professor Dragoslav Šiljak, whose hard work and dedication set a high standard for my professional development and growth.

This monograph also contains some results on decentralized control of multiple dynamical systems. When defining strategies or controls for multiple dynamical systems interacting among themselves, either in optimal or Lyapunov sense, great inspirations came from the delicate and thoughtful work and collaboration with late Prof. George Leitmann. George ended up being one of the greatest inspirations and supporters of my work. I also learned intricate details of game theory from the late Prof. Arik Artavazdovich Melikyan, with whom I had the privilege to work.

My journey into mathematics started with a remarkable person, who I was blessed to meet as a teenager. Late Prof. Vladimir Mićić, my essential advisor in mathematics, a Ph.D. student of late Prof. Vladimir Zorich, enlightened me since my early age with his witty, modest, and carrying personality. Because of him, I changed my professional path that was set to be in biology and chemistry, to mathematics.

I hold deepest gratitude to my former Ph.D. students who have worked on problems related to the material presented in this monograph. Professor Shankar Deka worked on global conditions for stability and stabilization of long short-term memory recurrent neural networks using two different approaches, which are related and even go beyond the results presented in this monograph. Dr. Massinissa Amrouche worked on homotopy-based computation of nonzero equilibria of long short-term memory neural networks as well as on the characterization of the activation functions for deep neural networks. I am very grateful to Massi for helping me with the computation of nonzero equilibria for the set of difference equation models of long short-term memory neural networks, presented in this book.

Competing Interests The author has no competing interests to declare that are relevant to the content of this manuscript.

Contents

Introduction, Inspirations and Motivations 1

Time is one of the essential measures of our existence, yet still an intriguing phenomenon we are trying to fully understand and explain. There are multiple complexities in understanding the meaning and the role of time. One thing is time itself, and the other is our perception of it, where the difference between the two is blur. An example of our perception and its illustration is how the movies are created, as series of images being run in time sufficiently fast so our eyes cannot detect the switches between the images. In this case the creation and the perception are clearly two different things. This exemplifies that our experience of time is, as expected, related to and influenced by the available sensors to detect the world around us. Essentially, the true meaning of time and its purpose are still mysteries. But that set aside, for the technical purposes of this book we will bypass this intriguing discussion and treat time in a formal way. It will be taken simply as an independent variable, while avoiding all of its philosophical intricacies and implications. But this does not mean that the essence of time is in any way disregarded but rather ignored for the sake of not being able to do more, at least in the given context. By simply treating time as an independent variable we are essentially admitting it is independent and not governed by any laws or that these laws if they exist are unknown to us.

In technical terms, simple yet sufficient for our purposes, treated as an independent variable, time also has different interpretations. These differences stem from the choice of allowable values that time, as an independent variable, takes. In that respect, there are two commonly known treatments of time. The first one is to consider time as a continuously changing variable, and the other is to consider it as a discrete variable. In the first case, time takes a continuum of real number values within one (bounded or unbounded) interval or a multiple of intervals, while in the second case it takes discrete values. The main topic of our interest is to study dynamical systems as their dynamics evolve over time either independent or dependent on our capabilities and desires to control them. The two main

D. Stipanović, *Difference Equations and Machine Learning*, Synthesis Lectures on Mathematics & Statistics, https://doi.org/10.1007/978-3-032-00910-4_1

mathematical descriptions of dynamical systems, dependent on how time is treated, are differential equation [3, 11, 21] and difference equation [24, 42] formulations, respectively. Differential equations, as representations of dynamical systems in continuous-time, are technically based on the integration process [61, 62], as its inverse process. Integration is a way of calculating volumes and thus belongs to and is studied by the measure theory [9, 16, 27, 37] as a discipline in mathematics. But then interesting enough, the process of integration is based on computations of limits of discrete sequences, which leads to another relation between the continuous and discrete realms. Therefore, this relation is quite sophisticated and bidirectional. This is yet again an illustration how even when we think we successfully formulated a categorization, it turns out not to be as sharp as initially presented. This remark on contradiction is of general type and, as it will appear evident, is also relevant for the material presented in this monograph. The contradictions do appear in our technical constructions, varying from modeling to constructions of proofs. Therefore, they need to be expected, welcomed, utilized, and embraced.

Looking from the discrete time prospective, discrete-time dynamical systems have a very specific attributes themselves, which can be illustrated by noticing that one discrete-time dynamical system may encompass a family of very different continuous-time dynamic behaviors. This means that one discrete-time dynamical system represents or comprises multiple continuous-time dynamical systems and their wide range of behaviors, obtained by different discretizations of different continuous-time dynamical systems, where each discretization results in the same discrete-time system. For example, one discrete-time dynamical system technically may include both smooth and discontinuous continuous-time dynamical systems with quite different dynamic behaviors. In the spirit of contradictions, let us now explore the opposite direction, that is from continuous to discrete. One of the motivations to consider discontinuities and revolutionize modeling behaviors of dynamical systems and differential equations goes back to spiritual beliefs of Dmitri Fyodorovich Egorov [20, 26]. Contrary to smooth scenarios where changes appear gradually, discontinuous changes can be related to more drastic changes in human behaviors caused by making sudden decisions that govern them. This was also one of the main points in the exemplary work of one of Egorov's students, St. Pavel Florensky [23, 50]. Egorov's descendants ended up being some of the most prominent mathematicians revolutionizing the field of mathematics like Luzin, Kolmogorov, Pontryagin, Arnold, Krylov, and Filippov, to name just a few. Practically this group of mathematicians constituted a core of one of the most influential schools of mathematics housed at M. V. Lomonosov Moscow State University. Their contributions to various branches of mathematics, like measure theory, functional analysis, probability theory, and theory of dynamical systems and differential equations, shaped modern mathematics as we know it, where different concepts of discontinuity play important and instrumental roles. Interesting enough, this group of mathematicians managed to also influence the foundations of machine learning and artificial intelligence. Some notable examples are the representation formulations and decompositions of functions on bounded domains [38] with applications to neural networks [41] and the maximum/minimum principle [51] being used in developing the back-propagation algorithm [43]. Finally, discontinuities may also stem from the artificial sources like the utilization of how the information updates are available or delivered to continuous-time dynamical processes, that is, at discrete time instances [60]. Furthermore, discontinuities of various kinds like singularities were recognized to capture most fundamental characteristics of the continuous-time dynamical systems [46, 49].

One of the greatest challenges when dealing with dynamical systems is our inability to, in general, predict and control their behaviors. These behaviors are represented by trajectories or solutions of the corresponding models of dynamical systems. This inability is despite the fact that the modeling itself can be done pretty accurately using either differential of difference equations. In line with this, one of the goals of this book is also to emphasize our inability to characterize, in technical terms, sophisticated behaviors of dynamical systems. For example, textbooks of Erich Kamke are still very complete in terms of existing closed-form solutions to differential equations [35, 36], where there are cases in which closed-form solutions do not even exist. Maybe not the right moment to start this discussion, yet why not at least state that we may be at the brink of a need for new mathematical formulations. Maybe a scientific revolution like the one sparked by Nikolai Ivanovich Lobachevsky is needed [48]. The problem is that for enabling and nurturing the atmosphere to support such radical inventions we need time to be spent and devoted without any promise of a successful outcome, which nowadays does not seem to be a popular idea. Nowadays, it seems like we rather choose numerical schemes enabled by power processors that provide numerical solutions that often fail to provide much of any deeper understanding and vision. It seems that the nature of the results follows the nature of how the tools were developed, that is, either by sophistication or by a brutal force. But here we have to stress that no one is disputing the essential role and relevance of numerical schemes [58]. Along those lines, one of the greatest inspirations of my overall work came from the sophisticated and brilliant theory of dynamical systems developed by Professor Aleksandr Mikhailovich Lyapunov [33, 34, 47]. This theory enabled us to provide guaranties for behaviors of general dynamical systems without having to compute or know explicitly their solutions. The brilliance of this mathematical theory is that it provides, in technical and rigorous ways, estimations of general dynamical systems' behaviors without a need to compute particular solutions. On one hand this is a sophisticated theory yet on the other it is of enormous application value. It is a general theory but its particular nature stems from the fact that for the general theory to be successfully applied, a particular Lyapunov function candidate has to be constructed to match the corresponding dynamical system, which behaviors are being investigated. This powerful theory also provides a way to control behaviors of dynamical systems. But again the approach has both general and particular natures. One of its main advantages when, for example, compared to optimal control [44] is that the control laws based on the Lyapunov function approach [25] can be computed in much simpler and computationally efficient forms, which for nonlinear systems affine in control can be computed even in the explicit, that is, closed forms. Nonlinear systems affine in control are an important class of dynamical systems since they represent realistic models for a wide variety of vehicles. In the case of multiple dynamical systems, this becomes even more evident since the multi-criteria problems involving multiple dynamical systems considered in the optimal sense are nothing but multi-player dynamical games [32, 49, 57]. Computation of optimal control laws or strategies for the players involve solving partial differential equations that are known for highly discontinuous solutions and the numerical computations scale very poorly with the problem dimension. In other words, the computations suffer from the so called "curse of dimensionality." Lyapunov formulation, on the other hand enables computation of closed-form control strategies for players in multi-player dynamical games for a substantial class of scenarios, with guaranteed outcomes.

There are a few technical motivations for the specific topics covered in this book. One goes back to my undergraduate engineering diploma thesis work on training neural networks to model and control discrete-time systems using the internal model control structure [55]. Then, a simple modified internal model framework was developed and tested, which enables a generally unknown dynamical system to follow a wide variety of reference trajectories. The other technical motivation, is a more recent interest in studying dynamic properties of a class of recurrent neural networks used in machine learning, with quite interesting, intriguing, and up to this day resistant to establishing rigorous technical descriptions of their common behaviors. Being unable to characterize dynamic behaviors of these recurrent neural networks seems to be related to controlling unknown dynamical systems. Furthermore, a regulator or a tracking problem seems to be related to recognizing or classifying input signals. The analyses are placed in discrete-time domain as a consequence of the particular recurrent neural network dynamical models being difference equations. Discrete-time dynamical systems also have a number of convenient properties, when compared to their continuous-time counterparts. One is that there are much simpler and straightforward analyses of existence and uniqueness of their solutions. Furthermore, being discrete in nature, these systems are well suited for a nowadays digitally run world around us based on digital computes and processors. While their value and use can still be argued, analog computers are practically not being used anymore, even for specific scientific applications where there is no need for discretization. Even though discrete-time systems and difference equations are simpler to study and analyze in some aspects, as will be illustrated in this monograph, we are still facing the fact that we can solve only so much. The collection of results in this monograph, serves as a simple testament to that.

I have been introduced to long short-term memory recurrent neural networks [31] by Professor Boris Murmann. It was his initial suggestion and encouragement that sparked my interest to study their dynamic behaviors related to machine learning applications, and in general as a very specific set of difference equations. This investigation started by finding a correspondence between the autonomous (that is, without input signals) nonlinear long short-term memory neural network model and its linearization around the corresponding equilibria. This relationship was established by recalling that the recurrent neural networks were recognized to work well when operating "at the edge of chaos" [6]. Chaotic motions [4, 30, 46] are generally considered to be any motions of dynamical systems that do not belong to a set of standard motions. The correspondence between the nonlinear and linearized models was initially studied for a special case when the gate bias vector in the nonlinear autonomous long short-term memory neural network model is set to zero, implying there is an equilibrium at the origin [56]. This simplification enabled an easy computation of the Jacobian of the linearized model. Numerically much more complicated scenario is when the equilibria are assumed to be nonzero, that is, the general case. This led to one particular consequence of studying dynamic behaviors of long short-term memory recurrent neural networks by linearization, which has been related to the computation of nonzero equilibria of their nonlinear difference equation models. These computations turn out to be computations of solutions of transcendental nonlinear algebraic equations that cannot be solved explicitly. Here an inspiration came from Professor Ljiljana Trajković and her work on homotopy theory and formulations, which we used for the equilibria computation. Related to the homotopy applications, I initially worked with my former Ph.D. student Dr. Massinissa Amrouche on homotopy based computation of nonzero equilibria of long short-term memory neural networks [1]. Massi also helped me with the computation of nonzero equilibria for the set of difference equation models of long-short-term memory neural networks, presented in this monograph.

Regarding recurrent neural networks, that is, neural networks modeled as dynamical systems, this monograph involves only a suitable set of research topics related to recurrent neural networks that I have been involved with. Many of these results have been a joint work with Professor Claire Jennifer Tomlin who not only worked with me but encouraged my most challenging attempts that I would not dare considering without her support. Other interesting and relevant research topics related to neural networks have been primarily led by my former Ph.D. students. Professor Shankar Deka formulated and proved global conditions for stability and stabilization of long short-term memory recurrent neural networks using two different approaches [13, 14]. Shankar also used sophisticated tools of nonlinear analysis in a characterization of adversarial perturbations and their implications in recurrent neural networks [15]. In addition to his earlier mentioned homotopy path-following computation of long-short term memory neural network equilibria work, Massi also worked on a characterization of activation functions for deep neural networks using partial differential equations [2]. This characterization, technically involved, provides a way to describe and explain the role of activation functions in deep neural networks.

This monograph also contains material on structured systems and decentralized control. With an increasing interest in controlling both natural and artificial complex dynamical systems, this problem also became of high practical value. Unfortunately the general formulation of the problem still remains as another technical problem we can solve only within very restrictive scenarios. A simple illustration is that even for linear time-invariant dynamical systems we are still lacking necessary and sufficient conditions for the existence of decentralized state feedback gain matrix structures, which is one of the simplest structured control problem formulations. Structured control [52, 53], as a concept, can be considered as a general problem formulation yet essentially it is not and thus should not be treated as such. The main reason why, is that the structured control will not work, in general, if there is no underlying matching structure of the dynamical system to be controlled. Thus, for the successful implementation and computation of a structured control law, its structure has to match the structure of the corresponding dynamical system and the computation should be feasible and efficient. One of the rare tools available for the computation of a certain class of structured controllers for a class of dynamical systems, is the convex optimization approach based on the linear matrix inequality formulation [7]. This approach, also used frequently in this monograph, can be applied to nonlinear systems which can be represented as a concatenation of its linearized part and a sector bounded nonlinearity. For a dynamical system modeled in this form, the linear matrix inequality optimization formulation is known to be efficient, even though only in a sufficient sense, for the computation of decentralized state feedback control laws. Some other structures, like decentralized overlapping structure where the states are shared among the subsystems, may be handled using linear matrix inequality optimization approach yet each scenario has to be taken as a particular problem with special attention. If the state feedback is replaced by the output feedback, the structured control problem becomes even more complicated to solve [54]. Overall, the linear matrix inequality optimization approach can be applied to some specific dynamical system formulations stemming from relevant applications, and in those cases given it is a convex optimization formulation, it is efficient and furthermore can be used to determine if the control problem is feasible. Analyzing and controlling dynamical systems involving structure, still remain challenging and in general terms unsolved. Structure itself has a discrete nature often characterized and captured by graphs [12]. Furthermore, there seems to be a lack of technical tools to understand and control the structural complexities, which are constantly evolving and morphing following the developments of more advanced artificial systems and our needs and tendencies to understand functionalities of more sophisticated natural systems.

Writing a book of any kind requires certain talents that I do not claim nor, in my opinion, possess. But I have certainly been inspired by a variety of literary masterpieces, which tackled most sophisticated questions of our existence and meaning. This mastery is related to how the ideas and opinions are conveyed to the readers, regardless of how technical or topic related the manuscript is. The origins of even technical works are often difficult to explain and track, yet unavoidably dependent on our opinions and beliefs they may be, as illustrated in the case of Egorov, results of nontechnical motivations and beliefs. After all, mathematics is based on axioms of common sense, where common sense may be quite a relative term. Regardless of how much sense this makes, and in line of what has just been discussed, I find it important to mention some of my greatest inspirations in the literary world. Fyodor Mikhailovich Dostoevsky's work left a deep mark on me and I am grateful to him for studying the deepest corners of our spirituality and personalities [17, 18], as well as I am greatful to Mikhail Afanasyevich Bulgakov for questioning our standard notions of good and evil [8], Umberto Eco for putting what makes sense in prospective [19], and Hermann Hesse for searching for meaning in life [28, 29]. Another example of a mastery in writing is how Professor Ross Barmish managed to write a mathematical textbook without equation numbers [5], which has been so skillfully done that it may not be even noticed by the reader, if not alerted beforehand.

The presentation style is chosen to be simple and to provide technical details when needed and appropriate, and therefore not to burden or overwhelm the reader when not necessary. This means that within the definitions, lemmas, and theorems, that is, mathematical constructions and statements, all technical details are provided while in the remaining textual content they may be omitted to provide a smoother and more engaging presentation. I hope that the chosen balance would also allow readers to get acquainted with my thought process and motivations that sometimes go beyond technical arguments and often are contradictory. The abbreviations have not been used, nor the appendices. These are personal preferences of mine, and nothing good or bad about their usage should be stipulated. This is simply chosen so that the presentation is streamlined and does not require much flipping among the pages in search of a meaningful information flow. Best efforts have been made to avoid unnecessary repetitions of the material, yet some repetitions have been utilized to help with the presentation flow, when necessary and at hopefully minimized number of occasions.

The structural organization of the monograph is also simple. The monograph contains one introductory chapter and three parts with seven chapters. The first chapter, the one not included in any of the parts, is this one, that is, the introduction. The first part contains material on basic properties of discrete-time dynamical systems and difference equations. Its first chapter provides some basic properties of difference equations and corresponding dynamical discrete-time systems, while its second chapter deals with the problem of stabilization of discrete-time dynamical systems. The second part is dedicated to the recurrent neural networks. The first chapter of the second part is related to dynamic properties of a special class of recurrent neural networks with quite intriguing behaviors as discrete-time dynamical systems. This class of recurrent neural networks is named long short-term memory neural networks [31] and it became well known in classification of time series. By now there is a number of new architectures that appear to be even more efficient in terms of classification of time series yet even they may integrate long short-term memory neural network units. The second chapter of the second part is dedicated to homotopy , that is, path-following formulations for the computation of nonzero equilibria of the long short-term memory neural networks. An exploration of utilizing convex combination property of gated recurrent unit neural networks [10] is provided in the third chapter of the second part. Gated recurrent unit neural networks represent a simplification of long short-term memory neural networks and in this monograph they serve as an illustration of how a convex combination formulation of dynamical system models can be utilized. Part three of this monograph is dedicated to a modified internal model control regulation of discrete-time dynamical systems. The first chapter of this part provides details on some simplifications and modifications of the original internal model control scheme. The second chapter of part three is a specification of the modified internal model control regulation for recurrent neural networks with bounded outputs due to the integration and utilization of sigmoid functions. Final remarks are summarized in the concluding thoughts section.

References

1. M. Amrouche, S. A. Deka, A. Lekić, V. R. Royo, D. M. Stipanović, B. Murmann, and C. J. Tomlin. Long Short-Term Memory Neural Network Equilibria Computation and Analysis. *Proceedings of the Workshop on Modeling and decision-making in the spatiotemporal domain, 32nd Conference on Neural Information Processing Systems (NIPS)*. Montreal, Canada, 2018.
2. M. Amrouche and D. M. Stipanović. A Formal Characterization of Activation Functions in Deep Neural Networks. *IEEE Transactions on Neural Networks and Learning Systems*, vol. 35, no. 2, pp. 2153–2166, 2014.
3. V. I. Arnold. *Ordinary Differential Equations*. Translated from the Third Edition in Russian by R. Cooke, Springer-Verlag, Berlin Heidelberg, 1992.
4. V. I. Arnold. *Catastrophe Theory*. Third, Revised and Expanded Edition, Translated from the Russian by G. S. Wassermann based on a Translation by R. K. Thomas, Second Printing, Springer-Verlag, Berlin Heidelberg, 2004.
5. R. B. Barmish. *New Tools for Robustness of Linear Systems*. Macmillan Publishing Company, New York, NY, 1994.
6. N. Bertschinger and T. Natschläger. Real-time computation at the edge of chaos in recurrent neural networks. *Neural Computation*, vol. 16, no. 7, pp. 1413–1436, 2004.
7. S. Boyd, L. E. Ghaoui, E. Feron, and V. Balakrishnan, *Linear Matrix Inequalities in System and Control Theory*, Philadelphia, PA, SIAM, 1994.
8. M. Bulgakov. *The Master and Margarita*. Vintage Books, New York, NY, 1996.
9. C. Carathéodory. *Algebraic Theory of Measure and Integration*. Second English Edition, Chelsea Publishing Company, New York, NY, 1963.
10. K. Cho, B. V. Merrienboer, C. Gulcehre, D. Bahdanau, F. Bougares, H. Schwenk, and Y. Bengio. Learning Phrase Representations using RNN Encoder-Decoder for Statistical Machine Translation. arXiv:1406.1078, 2014.
11. E. A. Coddington and N. Levinson. *Theory of Ordinary Differential Equations*. Mc-Graw Hill, New York, NY, 1955.
12. D. Cvetković. *Discrete Mathematical Structure*. In Serbian, Third Edition, Naučna Knjiga, Belgrade, Yugoslavia, 1987.
13. S. A. Deka, D. M. Stipanović, B. Murmann, and C. J. Tomlin. Global Asymptotic Stability and Stabilization of Long Short-Term Memory Neural Networks with Constant Weights and Biases. *Journal of Optimization Theory and Applications*, vol. 181, pp. 231-243, 2019.
14. S. A. Deka, D. M. Stipanović, B. Murmann, and C. J. Tomlin. Long Short-Term Memory Neural Network Stability and Stabilization using Linear Matrix Inequalities. *Proceedings of the 2019 IEEE International Symposium on Circuits and Systems*, Sapporo, Japan, 2019.
15. S. A. Deka, D. M. Stipanović and C. J. Tomlin. Dynamically Computing Adversarial Perturbations for Recurrent Neural Networks. *IEEE Transactions on Control Systems Technology*, vol. 30, no. 3, pp. 2615-2629, 2022.
16. J. L. Doob. *Measure Theory*. Graduate Texts in Mathematics, Springer-Verlag, New York Inc., 1994.

17. F. M. Dostoyevsky. *Crime and Punsihment*. In Serbian, "Jugoslavija" Publishing Company, Nemanjina 34, Belgrade, Yugoslavia, 1967.
18. F. Dostoyevsky. *The Brothers Karamazov*. Vintage Books, New York, NY, 1950.
19. U. Eco. *Foucault's Pendulum*. Ballantine Books, New York, NY, 1990.
20. D. Th. Egoroff. On Sequences of Measurable Functions. *Comptes rendus hebdomadaires des séances de l'Académie des sciences*, in French, vol. 152, pp. 244–246, 1911.
21. L. C. Evans. *Partial Differential Equations*, volume 19 of *Graduate Studies in Mathematics*. American Mathematical Society, Providence, RI, 2010.
22. A. F. Filippov. *Differential Equations with Discontinuous Righthand Sides*. Kluwer Academic Publishers, Dordrecht, The Netherlands, 1988.
23. P. Florensky. *The Pillar and Ground of the Truth*. Translated and Annotated by B. Jakim, Princeton University Press, Princeton, New Jersey, 1997.
24. G. F. Franklin, D. J. Powel, and M. L. Workman. *Digital Control of Dynamic Systems*. Addison-Wesley Longman, Inc., Menlo Park, California, 1998.
25. R. A. Freeman and P. V. Kokotović. *Robust Nonlinear Control Design: State Space and Lyapunov Techniques*. Birkhäuser, Boston, MA, 1996.
26. L. Graham and J.-M. Kantor. *Naming Infinity: A True Story of Religious Mysticism and Mathematical Creativity*. Harvard University Press, Belknap Press, 2009.
27. P. R. Halmos. *Measure Theory*. Graduate Texts in Mathematics, Springer-Verlag, New York Inc., 1974.
28. H. Hesse. *Steppenwolf*. H. Holt, New York, 1929.
29. H. Hesse. *Siddhartha*. H. Dover Publications, Mineola, NY, 2000.
30. M. W. Hirsch, S. Smale, and R. L. Devaney. *Differential Equations, Dynamical Systems, and an Introduction to Chaos*. Elsevier Inc., Waltham, MA, 3rd edition, 2013.
31. S. Hochreiter and J. Schmidhuber. Long short-term memory. *Neural Computation*, vol. 9, issue 8, pp. 1735–1780, 1997.
32. R. Isaacs. *Differential Games: A Mathematical Theory with Applications to Warfare and Pursuit, Control and Optimization*. John Wiley and Sons, Inc., New York, NY, 1965.
33. R. E. Kalman and J. E. Bertram. Control system analysis and design via the "second method of Lyapunov": I Continuous-time systems. *Journal of Basic Engineering, Transactions of the ASME*, vol. 82, pp. 371–393, 1960.
34. R. E. Kalman and J. E. Bertram. Control system analysis and design via the "second method of Lyapunov": II Discrete-time systems. *Journal of Basic Engineering, Transactions of the ASME*, vol. 82, pp. 394–400, 1960.
35. E. Kamke. *Differentialgleichungen: Lösungsmethoden und Lösungen I; Gewöhnliche Differentialgleichungen*. Leipzig, Germany, 1959. First edition was published in 1942.
36. E. Kamke. *Differentialgleichungen: Lösungsmethoden und Lösungen II; Partielle Differentialgleichungen 1: Ordnung für eine gesuchte Funktion*. Leipzig, Germany, 1944.
37. A. N. Kolmogorov and S. V. Fomin. *Introductory Real Analysis*. Translated and Edited by R. A. Silverman, Dover Publications, Inc., New York, 1975.
38. N. N. Kolmogorov. On the representation of continuous functions of many variables by superposition of continuous functions of one variable and addition. *Doklady Akademii Nauk SSSR*, 144:679–61, 1957; *American Mathematical Society Translation*, vol. 28, pp. 55–59, 1963.
39. N. N. Krasovskiĭ and A. I. Subbotin. *Game-Theoretical Control Problems*. Springer-Verlag, New York, NY, 1988.
40. N. V. Krylov. *An Introduction to the Theory of Diffusion Processes*. Translation of Mathematical Monographs, Volume 142, American Mathematical Society, 1995.
41. V. Kurková. Kolmogorov's Theorem and Multilayer Neural Networks. *Neural Networks*, vol. 5, pp. 501–506, 1992.

42. J. P. LaSalle. *The Stability and Control of Discrete Processes*. Springer-Verlag, New York, NY, 1986.
43. Y. LeCun. A Theoretical Framework for Back-Propagation. In D. Touretzky, G. Hinton, & T. Sejnowski (Eds.), *Proceedings of the 1988 Connectionist Models Summer School*, CMU, Pittsburgh, PA, pp. 21–28, 1988, Morgan Kaufmann.
44. G. Leitmann. *An Introduction to Optimal Control*. McGraw-Hill, In., New York, NY, 1966.
45. G. Leitmann. Guaranteed Avoidance Strategies. *Journal of Optimization Theory and Applications*, vol. 32, pp. 569–576, 1980.
46. G. A. Leonov. *Strange Attractors and Classical Stability Theory*. St. Petersburg University Press, St. Petersburg, Russia, 2008.
47. A. M. Liapunov. *The General Problem of the Stability of Motion*. Translated and edited by A. T. Fuller, Taylor & Francis, London, Washington, DC, 1992.
48. N. Lobachevski. *The Theory of Parallels*. Translated from the Original by G. B. Halsted, Open Court Publishing Company, La Salle, IL, 1914.
49. A. A. Melikyan. *Generalized Characteristics of First Order PDEs*. Birkhäuser, Boston, MA, 1998.
50. A. Oppo. *Antinomy and Symbol: Pavel Florensky's Philosophy of Discontinuity*. Koninklijke Brill BV, Leiden, The Netherlands, 2024.
51. L. S. Pontryagin, V. G. Boltyanski, R. V. Gamkrelidze, and E. F. Mishchenko. *The Mathematical Theory of Optimal Processes*. John Wiley & Sons, Inc., New York, NY, 1962.
52. D. D. Šiljak. *Large-scale Dynamic Systems: Stability and Structure*. North-Holland, New York, NY, 1978.
53. D. D. Šiljak. *Decentralized Control of Complex Systems*. Academic Press, Boston, MA, 1991.
54. S. S. Stanković, D. M. Stipanović and D. D. Šiljak. Decentralized Dynamic Output Feedback for Robust Stabilization of a Class of Nonlinear Interconnected Systems. *Automatica*, vol. 43, pp. 861–867, 2007.
55. D. M. Stipanović. *Controlling Nonlinear Systems using Neural Networks*. Diploma Thesis, Department of Automatic Control, School of Electrical Engineering, University of Belgrade, 1994.
56. D. M. Stipanović, B. Murmann, M. Causo, A. Lekić, V. R. Royo, C. J. Tomlin, E. Beigne, S. Thuries, M. Zarudniev, and S. Lesecq. Some local stability properties of an autonomous long short-term memory neural network model. *Proceedings of the 2018 IEEE International Symposium on Circuits and Systems*, 2018.
57. C. J. Tomlin, J. Lygeros and S. S. Sastry. *A Game Theoretic Approach to Controller Design for Hybrid Systems*. Proceedings of the IEEE 88 (7), 949-970, 2000.
58. D. D. Tošić. *Elements of Numerical Analysis*. In Serbian, Zavod za Udžbenike, Belgrade, Serbia, 2013.
59. Lj. Trajković. *Homotopy Methods for Computing DC Operating Points*. Encyclopedia of Electrical and Electronics Engineering, J. G. Webster, Ed., John Wiley & Sons, New York, vol. 9, pp. 171–176, 1999.
60. A. Zatezalo and D. M. Stipanović. Control of Dynamical Systems with Discrete and Uncertain Observations. *Discrete and Continuous Dynamical Systems-Series A*, vol. 35, pp. 4665–4681, 2015.
61. V. A. Zorich. *Mathematical Analysis I*. Springer-Verlag, Berlin Heidelberg, 2004.
62. V. A. Zorich. *Mathematical Analysis II*. Springer-Verlag, Berlin Heidelberg, 2004.

Part I
Discrete-Time Systems and Difference Equations

Difference Equations, Sequences and Discrete-Time Dynamical Systems

2

To begin the technical part of this monograph, let us recall and discuss some typical behaviors of difference equations [1–3, 5, 10, 26, 27, 32] and then provide ways of analyzing and characterizing these behaviors, to the extent possible of course. Difference equations model discrete-time dynamical systems where the independent variable, denoting progress in time, is discretized while the dependent variables are not quantized. When dependent variables are quantized, we refer to those dynamical systems as digital [13]. Throughout the book, we will consider difference equations in the following, quite general, vector form:

$$x(k+1) = f(k, x(k), u(k)) \tag{2.1}$$

where variable $k \in \mathbb{Z}^+$ take integer values and represents changing time instances and $\mathbb{Z}^+$ denotes the set of nonnegative integers. Furthermore, in Eq. (2.1), $f : \mathbb{Z}^+ \times \mathbb{R}^n \times \mathbb{R}^m \to \mathbb{R}^n$ is a vector function describing dynamics of the corresponding difference equation iterations, which for the purposes of this book is assumed to be continuous, if not stated otherwise. Vector variable $x(\cdot)$ is the state vector variable of dimension n, and $u(\cdot)$ is the control vector variable of dimension m. If not stated otherwise and without loss of generality, the initial time will be assumed to be the zeroth instance, and then the initial condition is simply denoted as $x(0)$. The state equation dynamics may be appended with the static output map, which is an algebraic equation commonly taking the following form:

$$y(k) = h(k, x(k), u(k)) \tag{2.2}$$

where $h : \mathbb{Z}^+ \times \mathbb{R}^n \times \mathbb{R}^m \to \mathbb{R}^p$ is an output function used to compute the specific output $y(\cdot)$ of dimension p. This function, if again not stated otherwise, will also be assumed to be continuous. In general, we refer to Eqs. (2.1) and (2.2) as time-varying and non-autonomous . Time-varying notation indicates that $f(\cdot)$ and $h(\cdot)$ have k as an argument appearing explicitly, and non-autonomous we use to indicate that the input appears as an

D. Stipanović, *Difference Equations and Machine Learning*, Synthesis Lectures on Mathematics & Statistics, https://doi.org/10.1007/978-3-032-00910-4_2

argument on the righthand side of the corresponding equation. Therefore, a time-invariant vector difference equation would be of the form

$$x(k+1) = f(x(k), u(k)) \tag{2.3}$$

where time-invariant and autonomous version of a dynamical system in discrete-time, would be simply represented by

$$x(k+1) = f(x(k)). \tag{2.4}$$

According to this notation, a time-varying autonomous vector difference equation would take the following expression:

$$x(k+1) = f(k, x(k)). \tag{2.5}$$

Similarly, a time-varying autonomous output algebraic equation would be of the following form:

$$y(k) = h(k, x(k)). \tag{2.6}$$

Notice that, technically speaking, difference equations (2.1)–(2.6) in the mentioned format are simply iterations and do not, in general, represent progressions in real time. Variable k, taking nonnegative integer values, appears in these expressions while *per se* providing no information on the durations of time between the samples, that is, successive values of k. Therefore these are quite general formulations that enable us to study simply discrete-time systems, which can be either modeling processes in real-time or off-line sequences. For example, various numerical schemes formulated as algorithms may not run in real-time and can be treated as iterations, and most commonly we refer to them as sequences. This also establishes a natural and basic connection to mathematical tools developed for general sequences [12, 24, 25]. On the other hand, difference equations obtained by discretizing continuous-time dynamical systems are linked to real time yet time instances can still be enumerated with k as long as the information of how discretization is being performed, is properly integrated into the righthand sides of difference equations modeling these systems [17].

Most commonly, the goal of the control is to come up either with an open-loop control function

$$u(k) = \zeta(k) \tag{2.7}$$

or a closed-loop function

$$u(k) = \xi(k, x(k)) \tag{2.8}$$

which may or may not depend on k explicitly, as in Eqs. (2.7) and (2.8), respectively. For example, after the control law is substituted into a non-autonomous difference equation we get either a time-invariant or time-varying autonomous difference equation depending if the control law explicitly depends on variable k or not. The control law is designed with the goal that the resulting vector autonomous difference equation would meet desired specifications.

Some of the well-known desired behaviors are convergence to an equilibrium or tracking a time-varying sequence. For example, an equilibrium denoted as x_* of a vector time-invariant difference equation is its constant solution, that is,

$$x_* = f(x_*) \tag{2.9}$$

also known as a fixed point of the mapping $f(\cdot)$ [36]. Furthermore, $f(\cdot)$ may be such that the corresponding iteration sequence $x(k+1) = f(x(k))$ converges to x_* either locally or globally, in terms of the initial value $x(0)$. One way to establish convergence is to construct a Lyapunov function, which is a scalar positive definite function (either locally or globally depending on the scenario) with respect to the equilibrium point, that is,

$$V(x_*) = 0. \tag{2.10}$$

Again, function $V(x)$ is positive in some neighborhood of the equilibrium so that it achieves the minimum value at the equilibrium as imposed by Eq. (2.10). This construction results in the equilibrium detectability, meaning

$$V(x) = 0 \equiv x = x_* \tag{2.11}$$

in the corresponding neighborhood. This also means that the level set zero is a single point, that is, the equilibrium itself. Therefore, zero value of the Lyapunov function is a way of encoding the corresponding equilibrium within the neighborhood of interest.

The computation of an equilibrium point of the corresponding discrete-time system, that is, the fixed point of the mapping $f(\cdot)$ can be done in various ways where a few that belong to path following methods [36] will be discussed later in this monograph. A convergence of the system trajectories can be established by checking if the following expression:

$$\Delta V(x(k)) = V(x(k+1)) - V(x(k)) \tag{2.12}$$

is negative in some neighborhood of the equilibrium. If this expression is non-positive then a boundedness of the sequence may be established. The simple argument behind this, is that the values of the Lyapunov function will not increase so the values of the state variables will be bounded by a finite value level set of the Lyapunov function. For the time-invariant autonomous system this may be treated using an algebraic expression in terms of the state variable as

$$\Delta V(x) = V(f(x)) - V(x). \tag{2.13}$$

To this inequality we can relate a control Lyapunov design [14] that would be catered towards convergence to the equilibrium point, given that the equation is time independent. In the case of time varying systems, this could be simply reformulated as

$$u(k, x) \equiv \arg\min_{u \in \mathbb{U}}(V(f(k, x, u)) - V(x)) \tag{2.14}$$

where $\mathbb{U}$ is the image set of the admissible control input, consisting of all values that the control variable can take. It is obvious that this formulation would be meaningful depending on the existence of the minimum on the righthand side of the expression in (2.14), which may be sometimes feasible in closed-forms yet, in general, has to be numerically solved. Of course, there must be an essential assumption that the minimum does exist, that is, that the problem is feasible. If the minimum exists, and can be solved explicitly, then the minimization operation would result in a closed-loop control law that does depend on the state variable and the iteration variable k. If the dynamics is time-invariant, that is, does not depend on the variable k explicitly, nor does the Lyapunov function, then the resulting control law would be so too, that is,

$$u(x) \equiv \arg\min_{u \in \mathbb{U}} (V(f(x, u)) - V(x)). \tag{2.15}$$

Since $V(\cdot)$ does not, in general (in some special applications it may), depend explicitly on u, and the minimization is done over the image set of u and not over the class of admissible control functions, which is very important to stress, this design simplifies to

$$u(x) \equiv \arg\min_{u \in \mathbb{U}} V(f(x, u)). \tag{2.16}$$

The minimization over the image set as in Eq. (2.16), and not over the class of admissible functions resembles the minimization process in Hamilton-Jacobi partial differential equations related to optimal control [11] and differential games [20]. As an illustration of the control Lyapunov design [14], let us consider a discrete-time dynamical system that is time-invariant and affine in control, that is, its dynamics is governed by the following righthand side:

$$f(x, u) = g(x) + h(x)u, \tag{2.17}$$

where we assume that $g(0) = 0$ and the goal is to stabilize this system with respect to the origin, as an equilibrium. We also assume that $h(x)$ in Eq. (2.17), never equals zero. If there is no prior knowledge which functional form of the Lyapunov function works for the system, one of the natural candidates to try first is the quadratic form [3], that is,

$$V(x) = x^T P x, \quad P \succ 0. \tag{2.18}$$

Here, square $n \times n$ matrix P is assumed to be positive definite, denoted as $P \succ 0$. The quadratic form given in Eq. (2.18), is an appropriate Lyapunov candidate to analyze system's behavior with respect to an equilibrium at the origin. Given this Lyapunov functional form, we can compute the following:

$$\begin{aligned} V(f(x, u)) &= f^T(x, u) P f(x, u) \\ &= u^T h^T(x) P h(x) u + 2 g^T(x) P h(x) u \\ &\quad + g^T(x) P g(x) \end{aligned} \tag{2.19}$$

where $h^T(x)Ph(x)$ is nonzero for every x, given our assumption that $h(x)$ never equals zero and that P is positive definite. We also assume that the control variable can take any vector value in its image set $\mathbb{U} = \mathbb{R}^m$. Then, this Lyapunov function candidate $V(x) = x^T Px$, which is again positive definite with respect to the equilibrium at the origin, results in $V(f(x, u))$ being a quadratic form in u and the resulting minimizer feedback control law would be the following:

$$u(x) = -(h^T(x)Ph(x))^{-1}h^T(x)Pg(x). \tag{2.20}$$

By substituting this feedback control law into the Lyapunov expression for the iterations in Eq. (2.19), we obtain

$$V(f(x, u(x))) = -g^T(x)Ph(x)(h^T(x)Ph(x))^{-1}h^T(x)Pg(x) + g^T(x)Pg(x). \tag{2.21}$$

Next, we can compute the Lyapunov difference, that is how the Lyapunov function values change along the iterations of the difference equation, as follows:

$$\begin{aligned} V(f(x, u(x))) - V(x) = & -g^T(x)(Ph(x)(h^T(x)Ph(x))^{-1}h^T(x)P - P)g(x) \\ & -x^T PX. \end{aligned} \tag{2.22}$$

Now, if the expression in Eq. (2.22), is negative definite in some neighborhood of the origin we can establish asymptotic stability using standard Lyapunov arguments.

An interesting special case of just presented derivation, is the case when $g(x) = Ax$ and $h(x) = B \neq 0$. Here A is an $n \times n$ matrix and B is an $n \times m$ matrix, by construction. In other words, we assume that the discrete-time system is linear and time-invariant. Then,

$$u(x) = -(B^T PB)^{-1}B^T PAx. \tag{2.23}$$

and

$$V(f(x, u(x))) - V(x) = -x^T(A^T PB(B^T PB)^{-1}B^T PA - A^T PA + P)x. \tag{2.24}$$

Therefore, if the following quadratic form related to Eq. (2.24):

$$A^T PB(B^T PB)^{-1}B^T PA - A^T PA + P \tag{2.25}$$

is negative definite, we can conclude global asymptotic stability of the system with respect to the origin. More about the formal Lyapunov stability definitions [21–23] and the computation of stabilizing control laws [18, 34] will be provided in the sections and chapters that follow. Especially complex is the problem of computing stabilizing control laws. The example of a closed-loop control law provided in Eq. (2.23) is a very singular and rare case when this is doable. More general is the scenario of controlling dynamical systems decomposed into a known linear dynamics part and an unknown but norm bounded nonlinear dynamics part. In that case convex optimization algorithms [8] based on linear matrix inequality formulations [8, 15] can be utilized. Linear matrix inequality formulations, though, are not general formulations and fit well scenarios where a quadratic Lyapunov function is used, and the constraints on the unknown linear part can also be represented as constraints on quadratic forms [35].

2.1 Convergence of Discrete-Time Dynamical Systems, Difference Equations, and Sequences

Discrete-time dynamical systems experience a range of different behaviors. When we discuss different behaviors of discrete-time dynamical systems, we consider those with no explicit appearance of input variables, that is, the autonomous ones. Therefore, they may be time-invariant or time-varying, but autonomous. These behaviors are essentially behaviors of the corresponding difference equations and can be viewed in relation to the independent variable representing time instances or just simply as iterations. This means that the independent variable taking integer values may be related to time or simply representing iterations in a sequence.

We will start this discussion by mentioning some of the typical behaviors, their motivations, and provide technical details in terms of their characterizations. The first behavior to mention, the one we already discussed, is the existence of and the systems' behavior with respect to equilibria related to solutions of the algebraic equation $x_* = f(x_*)$, that is, fixed points of $f(\cdot)$. These solutions may sometimes be computable in closed-forms yet more often have to be computed numerically. The defining equilibrium equation also indicates that x_* is a constant solution of the corresponding time-invariant difference equation. Equilibria can be defined in a similar way for time-varying difference equations as follows:

$$x_* = f(k, x_*), \ \forall k \in \mathbb{Z}^+, \tag{2.26}$$

where they are obviously fixed points for the mappings $f(k, \cdot)$ for each integer value of k, as Eq. (2.26) indicates. This means that in the time-varying case an equilibrium has to be a fixed point for a family of functions. Again, an equilibrium is on its own a constant solution of the corresponding difference equation and the system itself can have a number of different behaviors in the vicinity of these constant solutions. In order to provide a proper formulation of these characterizations of the behaviors, we need to do it in terms of the corresponding solutions of the system. Thus, let us denote a general solution of a discrete-time system, whether time varying or time-invariant, as $x(k; x_0; k_0$ where $x(k_0) = x_0$ is the initial condition at initial time $k_0 \in \mathbb{Z}^+$. Now, we are ready to define some standard stability behaviors in the sense of Lyapunov, which are comprised in the following definition [22].

Definition 2.1 An equilibrium point x_* of a discrete-time dynamical system is stable (or uniformly stable), in the sense of Lyapunov, if for any $\epsilon > 0$ there exist $\delta(\epsilon, k_0)$ (or $\delta(\epsilon)$) such that $\|x_0 - x_*\| \leq \delta$ implies $\|x(k; x_0, k_0) - x_*\| \leq \epsilon$ for all $k \geq k_0$.

In the case when trajectories of the system converge to an equilibrium, then the following concept of asymptotic stability applies, again in the sense of Lyapunov.

Definition 2.2 An equilibrium state x_* of a discrete-time dynamical system is asymptotically stable (or uniformly asymptotically stable), in the sense of Lyapunov, if it is stable (uniformly stable) and for any $\epsilon > 0$ there exist $T(\epsilon, k_0)$ (or $T(\epsilon)$) and $r(k_0) > 0$ ($r > 0$) such that $\|x_0 - x_*\| \leq r$ implies $\|x(k; x_0, k_0) - x_*\| \leq \epsilon$ for all $k \geq k_0 + T$.

Definitions 2.1 and 2.2 are all local concepts of equilibria stability. That is, they characterize behaviors of difference equations or discrete-time dynamical systems in the vicinity of the equilibria. If these results hold not only in some vicinities of the equilibria, but in the whole space, then global Lyapunov stability characterizations can be utilized. Global asymptotic equilibria stability formulations, in the sense of Lyapunov, are provided in the following definition.

Definition 2.3 An equilibrium state x_* of a discrete-time dynamical system is globally asymptotically stable (or globally uniformly asymptotically stable), in the sense of Lyapunov, if it is stable (uniformly stable) and all trajectories $x(k; x_0, k_0)$ are bounded (uniformly bounded) and converge to x_* for arbitrarily large $r > 0$.

Here we note that the uniform stability characterizations are well suited for those time-varying dynamic systems that essentially behave as time-invariant, at least from the stability point of view. Trivially, global asymptotic stability of any kind implies that the system has a unique equilibrium.

Testing conditions of the stability definitions in the sense of Lyapunov, as such, is quite difficult. What is much more convenient and feasible to do is to test, in general, sufficient conditions formulated in terms of Lyapunov functions. One of the key ingredients in providing a link between Lyapunov stability definitions and, in general, sufficient conditions based on Lyapunov functions, are special functions known as class $\mathcal{K}$ functions. This class of functions is defined as follows.

Definition 2.4 A scalar function $\alpha : [0, a) \to [0, +\infty)$ belongs to class $\mathcal{K}$ if $\alpha(0) = 0$ and if it a strictly increasing continuous function. It belongs to $\mathcal{K}_\infty$ class functions if it is a $\mathcal{K}$-function, $a = +\infty$ and $\lim_{r\to\infty} \alpha(r) = \infty$.

Class $\mathcal{K}$ functions play an instrumental role in connecting Lyapunov functions to the corresponding Lyapunov stability definitions, which, in the general setup, is done in one direction since the conditions are most often sufficient ones. Since the stability equilibria definitions and related conditions are commonly catered with an equilibrium at the origin in mind, let us recall a simple fact that a change of variables $z = x - x_*$ transforms any equilibrium into the origin and thus the analysis with respect to the origin as an equilibrium is of general nature.

Now, we can turn our attention to Lyapunov functions and assume that we have at our disposal a positive definite function $V(\cdot,\cdot) : \mathbb{R}^+ \times \mathbb{R}^n \to \mathbb{R}^+$. In technical terms, this means that there exists a class $\mathcal{K}$ function $\alpha(\cdot)$ such that

$$\alpha(\|x\|) \leq V(k, x), \tag{2.27}$$

which means that any positive definite function can be lower bounded by a class $\mathcal{K}$ function having a norm of the state variable as the argument.

Furthermore, we say that $V(\cdot,\cdot) : \mathbb{R}^+ \times \mathbb{R}^n \to \mathbb{R}^+$ is a decrescent function if there exist a class $\mathcal{K}$ function $\beta(\cdot)$ such that

$$V(k, x) \leq \beta(\|x\|). \tag{2.28}$$

Function $V(\cdot,\cdot)$ is called strictly decreasing, with respect to discrete-time dynamics $x(k+1) = f(k, x(k))$, if

$$V(k+1, f(k, x)) - V(k, x) \leq -\gamma(\|x\|),\ \forall k, \tag{2.29}$$

where $\gamma(\cdot)$ is another class $\mathcal{K}$ function. It is referred to as nonincreasing if

$$V(k+1, f(k, x)) - V(k, x) \leq 0,\ \forall k. \tag{2.30}$$

And finally it is called radially unbounded if $\alpha(\cdot)$ is a $\mathcal{K}_\infty$-function.

Now, a few Lyapunov function formulations for stability of discrete-time autonomous dynamical systems are formulated in the following theorem [21, 22]. Keep in mind, as mentioned earlier, that a Lyapunov stability analysis with respect to any nonzero equilibrium can be transformed to the analysis with respect to an equilibrium at the origin, by a simple change of variables. Thus, there is no loss of generality in formulating the following theorem with respect to an equilibrium at the origin.

Theorem 2.1 *An equilibrium at the origin is*

- *stable if the Lyapunov function is positive definite and nonincreasing;*
- *uniformly stable if the Lyapunov function is positive definite, decrescent, and nonincreasing;*
- *asymptotically stable if the Lyapunov function is positive definite and strictly decreasing;*
- *uniformly asymptotically stable if the Lyapunov function is positive definite, decrescent and strictly decreasing;*
- *asymptotically stable in the large if the Lyapunov function is positive definite, strictly decreasing and radially unbounded.*

A simple observation that Lyapunov functions are nonnegative and scalar by construction, leads to their connection to properties of nonnegative sequences. Yet here we can stress that one of the essential properties of the Lyapunov functions is that their minimum values are achieved at the corresponding equilibria, and not the non-negativeness *per se* but a special formulation where their minimal values serve as the *equilibria detectors*. But for the purpose of connecting Lyapunov functions to nonnegative sequences this property comes useful. Some ways to prove the convergence of these nonnegative sequences are related to difference inequalities. Now, let us simplify the notation in the following way:

$$\upsilon(k) \triangleq V(k, x(k)) \tag{2.31}$$

and compute an upper-bounding difference inequality

$$\upsilon(k+1) \leq \phi_k(\upsilon(k)). \tag{2.32}$$

One very relevant result that provides convergence of the sequences using a particular over-bounding is formulated in the following lemma [30].

Lemma 2.1 *Assume that $\upsilon(k) \geq 0$, $\alpha_k \geq 0$ and $\beta_k \geq 0$ for all $k \in \mathbb{Z}^+$, and*

$$\upsilon(k+1) \leq (1+\alpha_k)\upsilon(k) + \beta_k, \; \forall k \in \mathbb{Z}^+. \tag{2.33}$$

If, furthermore, both sums $\sum_{k=0}^{\infty} \alpha_k$ and $\sum_{k=0}^{\infty} \beta_k$ are convergent, then $\lim_{k\to\infty} \upsilon(k) = \upsilon^* \geq 0$.

Here let us explain some specific notation used in relation to sequences. We use the subscript dependence on time instances to denote samples of independent variables. However, the time dependence within parentheses is dedicated to dependent variables. A more general lemma with less restrictive conditions, is formulated next [4].

Lemma 2.2 *Assume that $\upsilon(k)$, α_k, β_k, and γ_k are all nonnegative samples of the corresponding sequences such that*

$$\upsilon(k+1) \leq (1+\alpha_k)\upsilon(k) + \beta_k - \gamma_k, \; \forall k \in \mathbb{Z}^+. \tag{2.34}$$

If, in addition, $\sum_{k=0}^{\infty} \alpha_k$ and $\sum_{k=0}^{\infty} \beta_k$ are convergent sums, then $\sum_{k=0}^{\infty} \gamma_k$ is convergent and $\lim_{k\to\infty} \upsilon(k) = \upsilon^* \geq 0$.

Now, the case when the multiplier of the independent variable is smaller than one, is considered in a stability or convergent result formulated in the following lemma [12].

Lemma 2.3 *Assume that $\upsilon(k)$, α_k, β_k, and γ_k are all nonnegative samples of the corresponding sequences such that*

$$\upsilon(k+1) \leq (1-\alpha_k)\upsilon(k) + \beta_k - \gamma_k, \ \forall k \in \mathbb{Z}^+. \tag{2.35}$$

If, in addition, $\alpha_k \in [0, 1)]$ for all $k \in \mathbb{Z}^+$ and $\sum_{k=0}^{\infty} \beta_k$ is a convergent sum, then $\sum_{k=0}^{\infty} \gamma_k$ is convergent while $\upsilon(\cdot)$ is bounded and $\lim_{k\to\infty} \upsilon(k) = \upsilon^* \geq 0$.

These conditions for nonnegative sequences can be used to establish stability of discrete-time dynamical systems using Lyapunov functions, as long as the bounding of the Lyapunov iterations satisfies one of the provided formulations in Lemmas 2.1–2.3. Lyapunov approach is also a way of scalarizing problem formulations involving vector variables. In particular, and related to the material just presented, problems of stability and convergence conditions with respect to the vector state variables of difference equations can be scalarized in terms of the values of the corresponding Lyapunov functions.

Another interesting and important behavior of discrete-time systems is the periodic behavior related to the existence of periodic motions or cycles. An existence of a periodic motion defined as a cycle of length m for a time-invariant model $x(k+1) = f(x(k))$ would mean that there are m samples $\{x_1, \ldots, x_m\}$ satisfying the following:

$$x(k+1) = f(x(k)). k \in \{1, \ldots, m-1\}, \ x(1) = f(x(m)). \tag{2.36}$$

So, technically speaking, an equilibrium can be viewed as a periodic motion of length one. Stability or a local stability property with respect to a periodic motion or cycle, can be defined via the following distance operator, associated with a particular cycle of length m, and any norm $\|\cdot\|$ in $\mathbb{R}^n$:

$$\rho_m(x) = \min_{i\in\{1,\ldots,m\}} \|x - x_i\|. \tag{2.37}$$

Now, the stability with respect to a periodic motion or a cycle, can be formally defined as follows.

Definition 2.5 A cycle of length m is stable (or uniformly stable) if for any $\epsilon > 0$ there exists $\delta(\epsilon, k_0)$ (or $\delta(\epsilon)$) such that $\rho_m(x(k_0)) \leq \delta$ implies $\rho_m(x(k; x_0, k_0)) \leq \epsilon$ for all $k \geq k_0$.

The concept of asymptotic stability with respect to a periodic motion or a cycle, similar to the equilibrium case, is formulated in the following definition.

Definition 2.6 A cycle of length m is asymptotically stable (or uniformly asymptotically stable) if it is stable (uniformly stable) and $\lim_{k\to\infty} \rho_m(x(k)) = 0$.

Given that we are considering periodic motions for time-invariant systems, the concepts of regular and uniform stability practically coincide. Assuming that $f(\cdot)$ is a continuously differentiable function, let us denote the Jacobian with respect to each point $\{x_1, \ldots, x_m\}$ defining the cycle of length m in the ordered way $[x_1, \ldots, x_m]$, as

$$J(x_i) = \left.\frac{\partial f(x)}{\partial x}\right|_{x=x_i}, \quad x_i \in \{x_1, \ldots, x_m\}. \tag{2.38}$$

Now, we can formulate some simple criteria for asymptotic stability and instability of time-invariant systems, respectively, as in the following theorem [31].

Theorem 2.2 *Let* $M = \prod_{i=1}^{m} J(x_i)$ *and let its spectral radius be denoted as* $\mu(M) = \max_{i\in\{1,\ldots,n\}} |\mu_i|$ *where* μ_i *denotes the i-th eigenvalue of* M*. Then,* $\mu(M) < 1$ *implies asymptotic stability while* $\mu(M) > 1$ *implies instability of the cycle.*

Notice that these sufficient condition results do not depend on which point we choose as the first in defining the cycle. This is a consequence of the fact that both products AB and BA of two square matrices A and B, which are of the same dimension, have exactly the same eigenvalues.

In training of recurrent neural networks, which are represented by difference equations so within our scope of interest, it has been long recognized that an efficient training results in the network, when considered as an autonomous system, operating at the "edge of chaos" [6]. While this has been recognized in terms of network behaviors, the technical characterizations of the "edge of chaos" for different dynamics and setups, are still missing. The recurrent neural network outputs are often of sigmoidal type, meaning sigmoid functions are integrated in their discrete-time dynamics, so they map either regions or the whole space of $\mathbb{R}^n$ onto bounded regions. Therefore, their image sets are bounded. This implies that, if a sigmoid function is used at the output of a sequence, no matter what the initial point is, starting with the second iteration the sequence mapping can be considered as a mapping between the same output image bounded set. For the scalar case, given a function that maps a bounded interval of $\mathbb{R}$ onto itself, there is the famous theorem of Sharkovskiĭ [33] and a theorem well known as for the title of the paper where it was published, that is, "period three implies chaos" [29], establishing very interesting properties of cycles and chaotic motions. Along the same line, is the following definition for chaos for the scalar case [17].

Definition 2.7 A map $f : [a, b] \to [a, b] \subset \mathbb{R}$ is chaotic if the following hold:

1. Periodic points of $f(\cdot)$ are dense in $\mathbb{I} = [a, b]$;
2. For any two sub-intervals $\mathbb{I}_1$ and $\mathbb{I}_2$ of $\mathbb{I}$, there is a point $x_0 \in \mathbb{I}_1$ and a positive integer $l0$ such that $f^l(x_0) \in \mathbb{I}_2$;

3. There is a constant ϵ such that, for any $x_0 \in \mathbb{I}$ and any open interval $\mathbb{U}$ around x_0, there is a point y_0 and a positive integer l, the inequality $|f^l(x_0) - f^l(y_0)| > \epsilon$ holds.

In general, chaotic behaviors are unpredictable and cannot be characterized as standard behaviors such as equilibria, stable, unstable, and periodic. Yet, for at least continuously differentiable maps that leave corresponding bounded sets invariant in terms of mapping, the "edge of chaos" may often be expected to be some kind of a periodic regime. This is only a stipulation that has been observed in the case of specific long short-term memory recurrent neural networks [19], as it will be discussed later in this monograph.

Related to cycles and chaos there are also relevant results on the topic that has been referred to as "stabilizing chaos." One of the interesting and convenient results refers again to the time-invariant discrete-time dynamical system $x(k+1) = f(x(k))$ where $f(\cdot)$ is assumed to be continuously differentiable. Furthermore, we assume that this system has a cycle of length m and that the cycle is unstable with an intention to stabilize it. Let us again, define matrix M as follows:

$$M = \prod_{i=1}^{m} J(x_i) \tag{2.39}$$

and let its spectral radius be denoted as $\mu(M)$ and defined as

$$\mu(M) = \max_{i \in \{1,\ldots,n\}} |\mu_i| \tag{2.40}$$

where μ_i denotes the i-th eigenvalue of M. The next theorem shows how to stabilize this cycle using a predictive control [31].

Theorem 2.3 *Assume that $f(\cdot)$ is continuously differentiable and that $x(k+1) = f(x(k))$ has a cycle of length m with $\mu(M) > 1$. Let μ_n such that $|\mu_n| = \mu(M)$ be real and $|\mu_i| < 1$ for $i \in \{1, \ldots, n-1\}$. Then, for p large enough, the following system*

$$x(k+1) = f(x(k)) - \epsilon(f^{m(p+1)+1}(x(k)) - f^{mp+1}(x(k)))$$

where

$$\frac{|\epsilon - \epsilon^*|}{|\epsilon^*|} < \frac{1}{\mu(M)^{1/m}}, \quad \epsilon^* = \frac{1}{\mu(M)^p(\mu(M) - 1)}$$

has the same cycle to be stable with respect to.

Let us turn again our attention to the time-varying difference iterations, that is, the following:

$$x(k+1) = f(k, x(k)), \; \forall k \in \mathbb{Z}^+$$

with initial condition $x(k_0) = x_0$. Here we assume that the initial time is k_0 rather than simply zero. Furthermore, let us again consider a general solution of this discrete-time dynamical system, denoted as $x(k, k_0, x_0)$. This solution, technically speaking, may be an equilibrium, that is, a constant solution or it may be a periodic solution, as examples of specific solutions. But in general it may be any solution, including chaotic solutions. For the remaining developments we only need to assume that $f(k, x)$ is continuously differentiable in x. The time varying Jacobian is then defined and can be computed as follows:

$$J(k, x) = \frac{\partial f(k, x)}{\partial x} \tag{2.41}$$

where the major issue is not the computation of the Jacobian itself, which is simply based on the partial derivative computation, but on the computation of the x argument of the Jacobian. Related to this issue, is the formulation of the linearized discrete-time dynamical model with respect to the solution $x(k, k_0, x_0)$, which can be written as

$$z(k+1) = J(k, x(k, k_0, x_0))z(k), \ \forall k \in \mathbb{Z}^+. \tag{2.42}$$

and it represents a dynamics of the system in the vicinity of the corresponding solution. Now, if this solution in Eq. (2.42), is for any reason available then the components of the linearized dynamical system are well defined. If on the other hand, the solution is not available and thus needs to be computed, like in the case of equilibria which may be solutions of the corresponding transcendental algebraic equations, we have to resort to often computationally expensive numerical methods. Regardless of the way the solution is provided, the fundamental matrix of the system $X(k, x_0)$ with the initial condition $X(k_0, x_0) = I_n$, where I_n denotes the $n \times n$ identity matrix, is defined and can be expressed as follows:

$$X(k, x_0) = \prod_{i=1}^{k-1} J(i, x(i, k_0, x_0)). \tag{2.43}$$

Now, the distance or discrepancy between the two solutions of nonlinear discrete-time dynamical system can be upper bounded and thus estimated as in the following lemma [28].

Lemma 2.4 *For any two solutions $x(k, k_0, x_0)$ and $x(k, k_0, y_0)$ starting at initial conditions x_0 and y_0, respectively, their distance over time can be estimated as follows:*

$$\|x(k, k_0, x_0) - x(k, k_0, y_0)\| \leq \sup_{v \in \mathcal{B}} \|X(k, v)\| \, \|x_0 - y_0\| \tag{2.44}$$

where $\mathcal{B} = \{v \mid \|x_0 - v\| \leq \|x_0 - y_0\|\}$.

Lemma 2.4 also shows the sensitivity of the system with respect to the initial conditions. More precisely, it shows the time evolving relationship, that is, the difference between two trajectories of the same system starting from two different initial conditions, which may or may not be close. This result can be extended to an open arcwise connected set Γ, where any two points in this set can be connected by a continuously differentiable map $\gamma(s) \in \Gamma$, where $s \in [0, 1]$. If we denote the length of this map by $\ell(\gamma, x_0, y_0)$ with $\gamma(0) = x_0$ and $\gamma(1) = y_0$, then the following estimate holds [28].

Lemma 2.5 *Assume that two solutions $x(k, k_0, x_0)$ and $x(k, k_0, y_0)$ start at initial conditions x_0 and y_0, respectively, belonging to an arcwise connected set Γ. Then,*

$$\|x(k, k_0, x_0) - x(k, k_0, y_0)\| \leq \sup_{v \in \Gamma} \|X(k, v)\| \, \ell(\gamma, x_0, y_0). \tag{2.45}$$

The estimation of a distance between two trajectories provided in Lemma 2.5 can be used to formulate stability conditions with respect to a particular dynamical system's solution, by taking one of these two trajectories to be the solution with respect to which we define stability. Along this line of thought, a discrete-time dynamical system is said to be globally stable with respect to the solution $x(k, k_0, x_0)$, if in addition to being Lyapunov stable it satisfies that for any initial condition $y_0 \in \mathbb{R}^n$ the following holds:

$$\lim_{t \to \infty} \|x(k, k_0, x_0) - x(k, k_0, y_0)\| = 0. \tag{2.46}$$

Now, the following global Lyapunov stability condition can be recalled [28].

Theorem 2.4 *Discrete-time system is globally stable with respect to the solution $x(k, k_0, x_0)$, if in addition to being Lyapunov stable the following holds:*

$$\lim_{t \to \infty} \sup_{v \in \mathbb{R}^n} \|X(k, v)\| = 0. \tag{2.47}$$

Opposite to stability, the system can also be unstable with respect to an equilibrium or, more generally, with respect to the solution $x(k, k_0, x_0)$. The most common definitions of instability behaviors simply state that the system is unstable if it is not stable, that is, if the stability conditions are violated. In practice, and as discussed earlier, the stability conditions can also be established using particular potential, that is, Lyapunov functions which are used to show that the trajectories are either converging to or staying bounded near nominal solutions being either equilibria or time-varying trajectories.

References

1. R. P. Agarwal and P. J. Y. Wong. *Advanced Topics in Difference Equations*. Kluwer Academic Publishers, Dordrecht, The Netherlands, 1997.
2. R. P. Agarwal. *Difference Equations and Inequalities: Theory, Methods, and Applications*. Second Edition, Revised and Expanded, Marcel Dekker, Inc., New York, 2000.
3. B. D. O. Anderson and J. B. Moore. *Optimal Control: Linear Quadratic Methods*. Prentice-Hall, Inc., Division of Simon and Schuster, One Lake Street, Upper Saddle River, NJ, 1990.
4. H. H. Bauschke and P. L. Combettes. *Convex Analysis and Monotone Operator Theory* . Second Edition, Springer International Publishing AG, Cham, Switzerland, 2017.

5. R. H. Bellman. *Methods of Nonlinear Analysis*. Volumes I and II, Academic Press, Inc., New York, NY, 1970.
6. N. Bertschinger and T. Natschläger. Real-time computation at the edge of chaos in recurrent neural networks. *Neural Computation*, vol. 16, no. 7, pp. 1413–1436, 2004.
7. S. Boyd, L. E. Ghaoui, E. Feron, and V. Balakrishnan. *Linear Matrix Inequalities in System and Control Theory*. SIAM, Philadelphia, PA, 1994.
8. S. Boyd and L. Vandenberghe. *Convex Optimization*. Cambridge University Press, 2004.
9. P. L. Combettes. 'Quasi-Fejérian Analysis of Some Optimization Algorithms. *Studies in Computational Mathematics*, vol. 8, pp. 115-152, 2001.
10. S. Elaydi. *An Introduction to Difference Equations*. Third Edition, Springer, New York, NY, 2005.
11. L. C. Evans. *Partial Differential Equations*, volume 19 of *Graduate Studies in Mathematics*. American Mathematical Society, Providence, RI, 2010.
12. B. Franci and S. Grammatico. Convergence of sequences: A survey. *Annual Reviews in Control*, vol. 53, pp. 161–186, 2022.
13. G. F. Franklin, D. J. Powel, and M. L. Workman. *Digital Control of Dynamic Systems*. Addison-Wesley Longman, Inc., Menlo Park, California, 1998.
14. R. A. Freeman and P. V. Kokotović. *Robust Nonlinear Control Design: State Space and Lyapunov Techniques*. Birkhäuser, Boston, MA, 1996.
15. P. Gahinet, A. Nemirovski, A. J. Laub, and M. Chilali. *LMI Control Toolbox: For Use with MATLAB*. The MathWorks, Natick, MA, 1995.
16. W. Hahn. *Theory and Application of Liapunov's Direct Method*. Prentice-Hall, Inc., Englewood Cliffs, NJ, 1963.
17. M. W. Hirsch, S. Smale, and R. L. Devaney. *Differential Equations, Dynamical Systems, and an Introduction to Chaos*, 3rd Edition., Elsevier Inc., Waltham, MA, 2013.
18. D. W. C. Ho and G. Lu. Robust stabilization for a class of discrete-time non-linear systems via output feedback: the unified LMI approach. *International Journal of Control*, vol. 76, no. 2, pp. 105–115, 2003.
19. S. Hochreiter and J. Schmidhuber. Long short-term memory. *Neural Computation*, vol. 9, issue 8, pp. 1735–1780, 1997.
20. R. Isaacs. *Differential Games: A Mathematical Theory with Applications to Warfare and Pursuit, Control and Optimization*. John Wiley and Sons, Inc., New York, NY, 1965.
21. R. E. Kalman and J. E. Bertram. Control system analysis and design via the "Second method of Lyapunov": I Continuous-time systems. *Journal of Basic Engineering, Transactions of the ASME*, vol. 82, pp. 371–393, 1960.
22. R. E. Kalman and J. E. Bertram. Control system analysis and design via the "Second method of Lyapunov": II Discrete-time systems. *Journal of Basic Engineering, Transactions of the ASME*, vol. 82, pp. 394–400, 1960.
23. H. K. Khalil. *Nonlinear Systems*, 3rd Edition, Prentice Hall, Upper Saddle River, NJ, 2002.
24. K. Knopp, *Theory and Application of Infinite Series*, 2nd Edition, Blackie & Son Limited, London and Glasgow, UK, 1954.
25. K. Knopp. *Infinite Sequences and Series*, Dover Publications, Inc., New York, NY, 1956.
26. V. Lakshmikantham and D. Trigiante. *Theory of Difference Equations: Numerical Methods and Applications*. Second Edition, Marcel Dekker, Inc., New York, NY, 1988.
27. J. P. LaSalle. *The Stability and Control of Discrete Processes*. Springer-Verlag, New York, NY, 1986.
28. G. A. Leonov. *Strange Attractors and Classical Stability Theory*. St. Petersburg University Press, St. Petersburg, Russia, 2008.
29. T.-Y. Li and J. A. Yorke. Period Three Implies Chaos. *The American Mathematical Monthly*, vol. 82, pp. 985–992, 1975.

30. B. T. Polyak. *Introduction to Optimization*, Optimization Software, Inc., Publications Division, New York, 1987.
31. B. T. Polyak. Stabilizing Chaos with Predictive Control. *Automation and Remote Control*, vol. 66, pp. 1791–1804, 2005.
32. H. Sedaghat. *Nonlinear Difference Equations: Theory with Applications to Social Science Models*. Springer, Dordrecht, The Netherlands, 2003.
33. A. N. Sharkovskiĭ. Coexistence of Cycles of a Continuous Map of the Line into Itself. *International Journal of Bifurcation and Chaos*, vol. 5, pp. 1263–1273, 1995. Originally published in *Ukrainian Mathematical Journal*, vol. 16, no. 1, pp. 61–71, 1964.
34. D. M. Stipanović and D. D. Šiljak. Robust stability and stabilization of discrete-time non-linear systems: The LMI approach. *International Journal of Control*, vol. 74, pp. 873–879, 2001.
35. V. A. Yakubovich. The S-procedure in nonlinear control theory. *Vestnik Leningradskogo Universiteta*, vol. 1, pp. 62–77, 1971, in Russian. English translation is in Vestnik Leningrad. Univ. Math, vol. 4, pp. 73-93.
36. W. Zangwill and C. Garcia. *Pathways to Solutions, Fixed Points, and Equilibria*. Englewood Cliffs, NJ, Prentice-Hall, 1981.

3 Stabilization of Difference Equations

Even though the concept of stabilization is very well known and technically formalized, we may stipulate that there is a lack of tools for computing stabilizing control laws. As already mentioned a few times in this monograph, one of the rare existing and efficient tools for the computation of control laws is based on convex optimization [5] and commonly known as the linear matrix inequalities formulation [4, 6]. It is important to note that this approach handles both stability and stabilization problem formulations yet these formulations are only for specific scenarios, which can be characterized as linear-like. By linear-like we mean classes of problems that can be linked to linear control theory and design where further details will be provided in what follows.

A general formulation of the linear matrix inequality formulation is to check if the following form:

$$\sum_{i=0}^{N} \gamma_i A_i \tag{3.1}$$

is either positive or negative definite or semidefinite. In expression (3.1), γ_i's are scalar variables, and A_i's are given square matrices of the same dimension. As a consequence of this formulation, any expression that contains matrix variables appearing linearly, if needed to be checked for its either negative or positive definiteness, is therefore a linear matrix inequality formulation. Convex programming formulation also allows to append a maximization or minimization of a convex, most commonly in a linear form, objective function. Then the problem of optimizing this linear convex function subject to any number of linear matrix inequality constrains is a convex problem. By doing so, it became common in the controls community to leave final results in terms of linear matrix inequality convex problems, given their convex nature. Convex optimization problems, if feasible, have global optimum solutions and a related convex algorithm is guaranteed to converge to it. So convex

D. Stipanović, *Difference Equations and Machine Learning*, Synthesis Lectures on Mathematics & Statistics, https://doi.org/10.1007/978-3-032-00910-4_3

programming can be used to determine whether the corresponding convex optimization problem is feasible.

Again, the linear matrix inequality optimization tool, while very efficient, can only be applied to specific classes of problems. To illustrate this, let us start with an autonomous time-varying discrete-time dynamical system and write it in the following form:

$$x(k+1) = Ax(k) + m(k, x(k) \tag{3.2}$$

where A is an $n \times n$ matrix and $m(k, x(k)$ is, in this context, assumed to be, and thus treated as, a perturbation. Notice that this form can be considered as a general form of the autonomous time-varying discrete-time system representation, by a simple observation that we can set

$$m(k, x(k)) = f(k, x(k)) - Ax(k). \tag{3.3}$$

where $f(k, x(k))$ represents a general form of the righthand side of difference Eq. (3.2). But then when we get to study dynamical properties of such a formulation we need to be more specific to be able to achieve any kind of accomplishment. For example, the assumption is that the linear part is stable and known while the perturbation is, in general unknown, and does not contribute to the system's stability. Trivial to claim is of course that if the perturbed term can be anything, there is no way to accurately estimate the system's dynamic behavior based on the linear part only, not locally nor globally. In line with this, we assume that the perturbation, as a function, satisfies the following "sector" condition:

$$m^T(k, x)m(k, x) \leq \alpha^2 x^T M^T M x \tag{3.4}$$

where a positive value of the parameter α reflects the size of the sector and matrix M of dimension $p \times n$ reflects its shape or structure. This assumption means that the perturbation is vanishing, that is,

$$m(k, 0) = 0, \ \forall k, \tag{3.5}$$

implying that an equilibrium of the original nonlinear system in Eq. (3.2), is at the origin. This also explains why we use "sector" to denote this type of bounding given its equivalent representation is the following:

$$\|m(k, x)\| \leq \alpha \|Mx\|, \ \alpha \geq 0. \tag{3.6}$$

Inequality (3.6) geometrically means that the perturbation is bounded by a linear function emanating from the origin. Here the norm represents the Euclidean norm.

Again, it is important to stress that the sector condition implies that the system has an equilibrium at the origin and that the stability and stabilization are considered in terms of the system's constant solution identically equal to zero. Now, we will illustrate the linear matrix inequality technique by investigating the stability of the overall system using a quadratic Lyapunov function [13–15, 17]

$$V(x) = x^T P x \tag{3.7}$$

for some positive definite matrix P. This Lyapunov function also fits the scenario where the system has an equilibrium at the origin, given its minimum is at the origin. This choice of Lyapunov function is also well suited for the nominal linear part but also for the sector bounding. A key ingredient in merging the two is the renowned *S-procedure* by Yakubovich [29]. It simply states that one quadratic form $Q_0(x)$ is negative definite under the condition that the other quadratic form $Q_1(x)$ is negative semidefinite if and only if there is a non-negative scalar τ so that $Q_0(x) - \tau Q_1(x)$ is negative definite. In technical terms this can be expressed as follows:

$$Q_0(x) \prec 0 \text{ when } Q_1(x) \preceq 0 \Leftrightarrow (\exists \tau \geq 0)(Q_0(x) - \tau Q_1(x) \prec 0). \tag{3.8}$$

It is important to note that in the case when there is exactly one semidefinite constraint as in expression (3.8), the *S-procedure* provides a necessary and sufficient relationship while in the case of more than one semidefinite condition it provides only a sufficient condition. To link the *S-procedure* expression (3.8) to our problem of interest, $Q_0(x)$ would represent the Lyapunov stability quadratic form and $Q_1(x)$ would represent the sector bounding condition on the nonlinearity. By applying *S-procedure*, and after some equivalent transformations, one obtains the following convex optimization problem [27]:

$$\begin{aligned} &\min \gamma \\ &\text{subject to} \\ &P \succ 0, \\ &\begin{bmatrix} A^T P A - P & A^T P & M^T \\ PA & P - I & 0 \\ M & 0 & -\gamma I \end{bmatrix} \prec 0, \end{aligned} \tag{3.9}$$

which, if feasible in P and γ, implies that the perturbed system in Eq. (3.2), is stable with the sector size

$$\alpha = 1/\sqrt{\gamma}. \tag{3.10}$$

Unfortunately, the linear matrix inequality formulation (3.9) is not appropriate for the control design because if the system matrix is changed to $A + BK$ then the problem, as is, is no longer linear in terms of the feedback gain matrix K. To recast and accommodate this formulation also for the control feedback design, we follow the approach proposed and formulated in [10]. The only difference is that we will derive explicitly the linear matrix inequality formulation for the static state feedback design. Technical details on more complicated designs, like output and dynamic controller designs can be found in the same reference, that is [10]. To fit the scenario, and also to end up with a feasible control design formulation, we will consider a nonlinear discrete-time time-varying and nonautonomous system in the following form:

$$\begin{aligned} x(k+1) &= Ax(k) + Bu(k) + m(k, x(k)) = (A + BK)x(k) + m(k, x(k)) \\ &= \hat{A}x(k) + m(k, x(k)) \end{aligned} \tag{3.11}$$

where $\hat{A}$ may represent both open and closed-loop matrix of the linear part. This means either $\hat{A} = A$ or $\hat{A} = A + BK$ where K represents a static feedback gain, which dimension is defined by dimensions of matrices A and B. Given that the nominal part is linear and time-invariant, we again choose the standard quadratic Lyapunov function $V(x) = x^T \hat{P} x$, with P positive definite, and follow derivations in [10] to compute the following:

$$\begin{aligned} V(x(k+1)) - V(x(k)) = & \ (\hat{A}x(k) + m(k, x(k))^T \hat{P}(\hat{A}x(k) + m(k, x(k)) \\ & -x(k)^T \hat{P} x(k) \\ = & \ x(k)^T (\hat{A}^T \hat{P} \hat{A} - \hat{P}) x(k) + m^T(k, x(k)) \hat{P} \hat{A} x(k) \\ & + x^T(k) \hat{A}^T \hat{P} m(k, x(k)) \\ & + m^T(k, x(k)) \hat{P} m(k, x(k)). \end{aligned} \quad (3.12)$$

Now, by treating $x(k)$ simply as an independent variable x and $m(k, x(k))$ as an independent variable m. Both the increment of Lyapunov function between the two successive instances, and the sector bound on the perturbation, can be treated as quadratic functions in terms of the joint variable vector $[x^T, m^T]^T$. Therefore, we can use the *S-procedure*, which results in the following expression:

$$x^T(\hat{A}^T \hat{P} \hat{A} - \hat{P})x + m^T \hat{P} \hat{A} x + x^T \hat{A}^T \hat{P} m + m^T \hat{P} m - \tau m^T m + \tau \alpha^2 x^T M^T M x. \quad (3.13)$$

Next, and without loss of generality we can impose that $\tau > 0$ and set

$$P = \hat{P}/\tau, \quad (3.14)$$

that is scale the Lyapunov function matrix while keeping its definiteness unaltered. Therefore, we obtain that the equivalent expression to the one in (3.13), in terms of checking whether it is negative definite, is the following:

$$x^T(\hat{A}^T P \hat{A} - P)x + m^T P \hat{A} x + x^T \hat{A}^T P m + m^T P m - m^T m + \alpha^2 x^T M^T M x. \quad (3.15)$$

This expression can be further rewritten in the following form:

$$\begin{aligned} & x^T(\hat{A}^T P \hat{A} - P + \alpha^2 M^T M + \hat{A}^T P (I - P)^{-1} P \hat{A}) x - x^T \hat{A}^T P (I - P)^{-1} P \hat{A} x \\ & + m^T P \hat{A} x + x^T \hat{A}^T P m + m^T (P - I) m + \alpha^2 x^T M^T M x \end{aligned} \quad (3.16)$$

where

$$\begin{aligned} & -x^T(\hat{A}^T P (I - P)^{-1} P \hat{A} - \alpha^2 M^T M) x + + m^T P \hat{A} x + x^T \hat{A}^T P m + m^T (P - I) m \\ & = -((I - P)^{-1/2} P \hat{A} x - (I - P)^{1/2} m)^T ((I - P)^{-1/2} P \hat{A} x - (I - P)^{1/2} m) \\ & = -\|(I - P)^{-1/2} P \hat{A} x - (I - P)^{1/2} m\|^2. \end{aligned} \quad (3.17)$$

Therefore, a sufficient condition for stability (or stabilization) under the assumed sector bound on the perturbation, would be to establish negative definiteness of the following matrix expression:

$$\begin{aligned}&\hat{A}^T P\hat{A} - P + \alpha^2 M^T M + \hat{A}^T P(I-P)^{-1}P\hat{A}\\&= -P + \alpha^2 M^T M + \hat{A}^T(P + P(I-P)^{-1}P)\hat{A}\\&= -P + \alpha^2 M^T M + \hat{A}^T(P^{-1} - I)^{-1}\hat{A}\end{aligned} \tag{3.18}$$

where we used the following identity:

$$P + P(I-P)^{-1}P = (P^{-1} - I)^{-1}. \tag{3.19}$$

A definiteness of any quadratic form will not be changed if multiplied from both sides by

$$Q = P^{-1} = Q^T, \tag{3.20}$$

given that matrix Q is both nonsingular and symmetric, by construction. From this point, matrix Q takes the role of the Lyapunov function. Now, by multiplying expression (3.18) from both sides by Q and using the change of variables in Eq. (3.20), results in its definiteness equivalent expression given as follows:

$$-Q + \alpha^2 QM^T MQ - Q\hat{A}^T(I-Q)^{-1}\hat{A}Q. \tag{3.21}$$

Now, we can use an equivalent transformation that is based on an application of the following well known result for structured matrices in a block form [4]:

$$\begin{bmatrix} R_{11} & R_{12}^T \\ R_{12} & R_{22} \end{bmatrix} \prec 0 \Leftrightarrow R_{22} \prec 0 \wedge R_{11} - R_{12}^T R_{22}^{-1} R_{12} \prec 0. \tag{3.22}$$

The equivalence provided in (3.22), can be used for our purpose by setting the following:

$$\begin{aligned}R_{11} &= -Q + \alpha^2 QM^T MQ,\\R_{12} &= \hat{A}Q,\\R_{22} &= I - Q.\end{aligned} \tag{3.23}$$

This choice of variables results in obtaining an equivalent, definiteness wise, matrix form of the expression in (3.21), as follows:

$$\begin{bmatrix} -Q + \alpha^2 QM^T MQ & Q\hat{A}^T \\ \hat{A}Q & I - Q \end{bmatrix} \tag{3.24}$$

By applying the same, negative or positive definiteness invariant, transformation, now with the following choice:

$$
\begin{aligned}
R_{11} &= \begin{bmatrix} -Q & Q\hat{A}^T \\ \hat{A}Q & I-Q \end{bmatrix}, \\
R_{12} &= \begin{bmatrix} MQ & 0 \end{bmatrix}, \\
R_{22} &= -\gamma I = -\alpha^{-2} I,
\end{aligned} \tag{3.25}
$$

we get the final matrix expression to be checked for its negative definiteness, as follows:

$$
\begin{bmatrix} -Q & Q\hat{A}^T & QM^T \\ \hat{A}Q & I-Q & 0 \\ MQ & 0 & -\gamma I \end{bmatrix}. \tag{3.26}
$$

Therefore, the linear matrix formulation of the combined stability and sector bound constraints accompanied with the maximization of the sector size, can be written as the following convex optimization problem:

$$
\begin{aligned}
&\min \gamma \\
&\text{subject to} \\
&Q \succ 0 \\
&\begin{bmatrix} -Q & Q\hat{A}^T & QM^T \\ \hat{A}Q & I-Q & 0 \\ MQ & 0 & -\gamma I \end{bmatrix} \prec 0.
\end{aligned} \tag{3.27}
$$

The optimization problem (3.27) would be an appropriate linear matrix optimization formulation if we are just interested in stability. If we are interested in computing a stabilizing gain matrix K, as incorporated in the following closed-loop system matrix expression:

$$
\hat{A} = A + BK \tag{3.28}
$$

then to end up with a proper linear matrix optimization formulation for stabilization, we also need to introduce the following change of variables:

$$
L = KQ. \tag{3.29}
$$

This enables us, to formulate the computation of the stabilizing controller for the overall system as the following convex optimization problem:

$$\begin{aligned}
&\min \gamma \\
&\text{subject to} \\
&Q \succ 0 \\
&\begin{bmatrix} -Q & QA^T + L^T B^T & QM^T \\ AQ + BL & I - Q & 0 \\ MQ & 0 & -\gamma I \end{bmatrix} \prec 0,
\end{aligned} \tag{3.30}$$

in terms of variables γ, Q, and L. If the convex optimization problem (3.30) is feasible, then the feedback control gain matrix is computed as

$$K = LQ^{-} = LP, \tag{3.31}$$

the sector size is computed as $\alpha = 1/\sqrt{\gamma}$, and the Lyapunov matrix is computed as $P = Q^{-1}$, where the positive scaling factor needs not to be included. So not only that this approach enables us to compute the parameters of the stabilizing controller (as being entries of the gain matrix K) but also to maximize the sector bound on the allowable vanishing perturbation. It is very important to stress again the value of the linear matrix inequality convex optimization formulation, which can be emphasized by repeating that before it became available as a tool, there was practically no efficient way for computing multi-variable state-space controllers. There were precise conditions for its existence but not efficient tools for its computation. If and when the approach is applicable, the feasibility of the problem is, as also mentioned earlier, uniquely determined by the convergence of the corresponding convex optimization algorithm. This makes the approach very convenient and efficient, again if applicable.

3.1 Vector Lyapunov Functions, Interconnected Systems and Connective Stability

So far we have not assumed any particular structure on the system dynamics nor on the control. Here by structure we mean the information about the interconnections among various parts of the system. Many dynamical systems created and built by humans or found in nature have particular structures, which could be explored and used to refine the analysis of these dynamical systems by incorporating the knowledge about these structure. Furthermore, the information about these structures is relevant to the control design when the goal is to control and regulate the corresponding dynamical systems. These implications and potential benefits may be in computing simpler, less computationally complex, easier to implement, and more reliable control designs for the structured systems. The motivations may also stem from simpler objectives, like our interests in maintaining and understanding the behaviors of structured dynamical systems. One of the basic structures is when an overall dynamical system can be represented as an interconnection of multiple lower dimensional subsystems. Often we refer to these systems as interconnected or simply complex dynamical systems.

A stability concept that fits this structure, is the concept based on so-called vector Lyapunov functions. Vector Lyapunov functions were independently introduced by Matrosov in [18] and Bellman in [3]. They are well suited for interconnected, that is, complex or even for what is nowadays also known and referred to as multi-agent systems. Interconnected dynamical systems also represent decentralized dynamical systems [23, 24]. The case when the subsystems share some common variables is referred to as decentralized overlapping structure, which nowadays in a broad sense is also referred to as a distributed structure.

Basic assumption for treating a dynamical system as an interconnected one, is an existence of N subsystems governed by their own dynamics. A quite general discrete-time dynamical system model formulation for interconnected systems would be as follows:

$$x_i(k+1) = f_i(k, x(k), u_i) = g_i(k, x_i(k), u_i) + h_i(k, x(k)), \;\; i \in \{1, ..., N\} \tag{3.32}$$

where x_i is the i-th subsystem's state vector and u_i is its input. Its "local" subsystem dynamics is captured by $g_i(k, x_i(k), u_i)$ and $h_i(k, x(k))$ is a concatenation of the interconnections between the i-th subsystem and the other subsystems. A concatenation of all local states is denoted by

$$x = [x_1^T, ..., x_N^T]^T. \tag{3.33}$$

In the compact form, the overall interconnected system (3.32) can be rewritten as

$$\dot{x} = f(t, x, u), \;\; \text{where } u = [u_1^T, ..., u_N^T]^T. \tag{3.34}$$

Again, the main idea of using Lyapunov theory is to use a scalar positive definite function to establish some insights, mainly stability and convergence properties, of a multi-variable dynamical system. The same general idea applies to the case of vector Lyapunov function approach where the overall, that is, composite Lyapunov function is constructed from a vector of individual subsystem Lyapunov function denoted as the following:

$$\nu(k, x) \equiv [v_1(k, x_1), ..., v_N(k, x_N)]^T. \tag{3.35}$$

The overall composite Lyapunov function is then computed as [25, 26, 28]

$$v(k, x) = \sum_{i=1}^{N} d_i v_i(k, x_i) = d^T \nu(k, x) \tag{3.36}$$

where, again, $v_i(\cdot)$ in Eq. (3.26) are positive definite individual subsystem functions. Positive weighting parameters denoted as d_i's are to be computed, as it will be shown to fit the system interconnections, that is, system's structure. The parameter vector d, which is positive element wise, is constructed from its elements being d_i's, which can also be viewed as weighting coefficients, in a simple way as follows:

$$d = [d_1, \ldots, d_N]^T.$$

Each individual subsystem Lyapunov function $v_i(k, x_i)$ is assumed to satisfy its local subsystem discrete-time Lyapunov inequality

$$\Delta_g v(k, x(k)) \equiv v_i(k+1, g_i(k, x_i, u_i)) - v_i(k, x_i) \leq -l_i w_i(k, x_i), \; i \in \{1, ..., N\}, \tag{3.37}$$

for each individual subsystem control law $u_i(\cdot)$, where $w_i(k, x_i)$ are positive definite functions in x_i. We also assume that local Lyapunov functions satisfy the triangle inequality, which is a very basic and standard assumption on Lyapunov functions. Therefore,

$$v_i(k, a+b) \leq v_i(k, a) + v_i(k, b), \;\; i \in \{1, ..., N\}, \forall k, \forall a, b \in \mathbb{R}^{n_i}. \tag{3.38}$$

In this framework, the interconnection part of each subsystem dynamics, denoted as $h_i(k, x(k))$ in Eq. (3.32), is treated as a perturbation to the system. Other approaches, where the interconnections are well known, may utilize interconnections in a more productive way, for example, to enhance the overall system's performance. In our approach, where again interconnections are treated as perturbations, if we do not impose some kind of bounds on the perturbations, there is nothing we can expect to be accomplished. In this setup, and also related to earlier introduced and motivated "sector" bounds, it is natural and feasible for the analysis to be developed, to assume the following "conical" type of bounding:

$$v_i(k+1, h_i(k, x)) \leq \sum_{j=1}^{N} b_{ij} w_j(k, x_j), \;\; i \in \{1, ..., N\}, \forall k \in \mathbb{Z}^+ \tag{3.39}$$

which also, by contraction, indicates that we are considering vanishing perturbations. Again, this bounding is in line with considering interconnections as unknown perturbations, except for their specific "conical" bounds which define areas in which the perturbations may vary. Coefficients b_{ij} in (3.39) need to be nonnegative with preferably as large values as possible since their larger values provide larger perturbation bounds resulting in guaranteeing more robustness to the system. As large as possible here is meant in the sense that the interconnection bounds are as large as possible while not ruining the system's stability.

Given the composite Lyapunov function in Eq. (3.36), we can compute the following Lyapunov increments along the overall system dynamics:

$$\begin{aligned} v(k+1, x(k+1)) - v(k, x(k)) &= \sum_{i=1}^{N} d_i (v_i(k+1, x(k+1)) - v_i(k, x(k))) \\ &\leq \sum_{i=1}^{N} d_i (v_i(k+1, g_i(k, x_i(k), u_i) + h_i(k, x(k))) \\ &- v_i(k, x(k))) \\ &\leq \sum_{i=1}^{N} d_i (\Delta_g v(k, x(k)) + v_i(k+1, h_i(k, x(k)))) \end{aligned}$$

$$\leq \sum_{i=1}^{N} d_i(-l_i w_i(k, x_i(k)) + \sum_{j=1}^{N} b_{ij} w_j(k, x_j(k)))$$
$$= -d^T Q w(k, x(k)) \tag{3.40}$$

where

$$w(k, x(k)) = [w_1(k, x_1(k), \ldots, w_N(k, x_N(k))]^T. \tag{3.41}$$

Matrix

$$Q = [q_{ij}] \tag{3.42}$$

is defined as an $N \times N$ matrix with elements

$$q_{ij} = \begin{cases} l_i - b_{ii}, & i = j \\ -b_{ij}, & i \neq j \end{cases} \tag{3.43}$$

The way matrix Q is constructed in Eqs. (3.40) and (3.43), implies that it is a square matrix with non-positive off-diagonal elements [2, 7, 11, 12]. If we consider the class of all square matrices with non-positive off-diagonal elements, then a very special subset of these matrices is known as the class of M-matrices [2]. They are very well known due to their connection to element by element nonnegative matrices and their relation to interconnected and quasi dominant dynamical systems. As it will be shown, what we need to establish stability of the interconnected discrete-time dynamical system with conical bounds on the interconnections, using composite Lyapunov function, are some specific properties of M-matrices. One of these properties is that if Q is an M-matrix [2] then there exist element-wise positive vectors d and c such that

$$d^T Q = c^T. \tag{3.44}$$

If satisfied, the condition in Eq. (3.44) would provide that the increments of the composite Lyapunov function are negative and thus the stability of the interconnected system will be established.

Given their importance for interconnected dynamical systems, let us at this point recall some characterizations of the set of M-matrices. A square matrix is an M-matrix if one of the equivalent conditions in the following theorem, is satisfied [23].

Theorem 3.1 *Assume that a square matrix $Q \in \mathbb{R}^{n \times n}$ has all off-diagonal elements non-positive, that is, $q_{ij} \leq 0$, $i \neq j$. Then, it is an M-matrix if one of the following equivalent conditions is satisfied:*

a. *There exists an element by element positive vector $d = [d_1, \ldots, d_n]^T > 0$ such that $Qd > 0$, that is, Qd is also an element by element positive vector.*
b. *There exists an element by element positive vector $d = [d_1, \ldots, d_n]^T > 0$ such that $Q^T d > 0$, that is, $Q^T d$ is also an element by element positive vector.*

c. *There exists a positive definite diagonal matrix* $D = \text{diag}\{d_1, \ldots, d_n\} \succ 0$ *such that* $Q^T D + DQ \succ 0$. *In other words, Q is positive stable with the corresponding Lyapunov matrix that can be computed in a diagonal form.*
d. *Matrix Q is nonsingular, that is, Q^{-1} exists and $Q^{-1} \geq 0$, so Q is a nonnegative matrix element by element.*
e. *Matrix Q is positive stable. This means that all eigenvalues of Q have positive real parts, that is,* $\text{Re}(\lambda_i(Q)) > 0$, $i = 1, \ldots, n$.

The knowledge of the properties of M-matrices, enables us to formulate the following theorem providing a sufficient condition that the decentralized control laws will stabilize the overall discrete-time interconnected dynamical system using the composite Lyapunov function, defined and constructed using the vector of subsystem Lyapunov functions.

Theorem 3.2 *If matrix Q is an M-matrix, then the interconnected system is stabilized with the decentralized control laws.*

Proof We already established in the derivations (3.40) that

$$v(k+1, x(k+1)) - v(k, x(k)) \leq -d^T Q w(k, x(k)) \tag{3.45}$$

where Q is an M-matrix. Then, there exist a positive vector c such that $c^T = d^T Q$ and therefore

$$v(k+1, x(k+1)) - v(k, x(k)) \leq -c^T w(k, x(k)) = \sum_{j=1}^{N} c_j w_j(k, x_j(k))). \tag{3.46}$$

The fact that the righthand side is negative definite in terms of the vector state variable x, completes the proof. □

Checking if Q is an M-matrix can be done using conditions of Theorem 3.1 and either linear programming or any convex programming package for linear matrix inequalities [4, 5], as shown in [25, 28], respectively. It is also interesting to point out that the composite function can be viewed as a scalarization of a multiobjective problem of stabilizing multiple subsystems, which is often a way to handle multiobjective problems [19]. This formulation can also be related to Pareto optimization formulations since a convex combination of multiple objectives corresponds to a Pareto optimal point with respect to the vector of individual subsystem Lyapunov functions treated as objective functions [21, 22].

The "conical" bounding of the interconnections can be utilized to cover more refined bounding structures. One of them is related to the concept of connective stability [24, 28], which is defined in terms of the interconnection functions. In particular, each interconnection function is assumed to be decomposable in the following form:

$$h_i(k, x) = \sum_{j=1}^{N} e_{ij}(k, x) h_{ij}(k, x) \tag{3.47}$$

where $e_{ij}(k, x)$'s represent scalar connective factors and $h_{ij}(k, x)$ represents the interconnection from subsystem j to subsystem i. The connective factors in Eq. (3.47) are assumed to be bounded and to satisfy the following:

$$e_{ij}(k, x) \in [0, \bar{e}_{ij}], \ \bar{e}_{ij} \geq 0. \tag{3.48}$$

Obviously, when a connection factor equals zero, this may be understood as the corresponding interconnection link being removed or being down. Therefore, the connection factors are bounded and we can concatenate them by assuming they are elements of an interconnection matrix $E(k, x)$ of dimension $N \times N$. The bounding parameter $\bar{e}_{ij}$ in Eq. (3.48) reflects how strong the interconnection from subsystem j to subsystem i, can be. The fundamental interconnection matrix $\bar{E}$ is defined as a square $N \times N$ matrix with entries $\bar{e}_{ij}$. Depending on the structure of the system, the fundamental interconnection matrices can take values in a box-like or a polytopic set. Given that a box can be represented as a polytope, we will assume the bounding set is a polytope and denote it as $\mathbb{E}$. Therefore, it can be represented as a convex hull of vertex matrices, that is, vertices in the space of matrices. We assume that there are ℓ vertex matrices denoted as $\bar{E}^i$ so that

$$\mathbb{E} = \text{conv}\{\bar{E}^i\}. \tag{3.49}$$

Now, we can define the set of allowable interconnections, through the set of allowable interconnection matrices, denoted as $\mathcal{E}$, as follows:

$$\mathcal{E} = \{E : \exists \bar{E} \in \mathbb{E}, \ |E| \leq \bar{E}\}. \tag{3.50}$$

For the structured interconnected system, we can now recall the connective stability definition as formulated in the following theorem [25, 28].

Definition 3.1 The structured interconnected system is connectively stable if it is stable in the sense of Lyapunov for any possible interconnection matrix $E \in \mathcal{E}$.

In order to be able to establish connective stability based on the assumed structured interconnections, we need to assume an appropriate bounding. First, let us utilize the triangle inequality property for the subsystem Lyapunov functions (3.38) to obtain the following:

$$v_i(k+1, h_i(k,x)) = v_i\left(k+1, \sum_{j=1}^{N} e_{ij}(k,x)h_{ij}(k,x)\right)$$
$$\leq \sum_{j=1}^{N} v_i\left(k+1, e_{ij}(k,x)h_{ij}(k,x)\right) \tag{3.51}$$

To get a proper upper bound, we assume that the positive definite functions $v_i(k, x_i)$ and $w_i(k, x_i)$, are all decrescent. Now, for each i we take their upper-bounding $\mathcal{K}$ functions and create one $\mathcal{K}$ function that over-bounds both of them. This can be done in a simple way by taking the maximum function point-wise, for example, as an over-bounding $\mathcal{K}$ function. That function, denoted as $\delta_i(\cdot)$, by construction is continuous and increasing, and satisfies the following:

$$v_i(k, x_i) \leq \delta_(\|x_i\|), \forall k,$$
$$w_i(k, x_i) \leq \delta_(\|x_i\|), \forall k, \tag{3.52}$$

where $\delta_i(\cdot)$ is a $\mathcal{K}$ function for each $i \in \{1, \ldots, N\}$. Therefore,

$$v_i(k+1, e_{ij}(k,x)h_{ij}(k,x)) \leq \delta_i(\|e_{ij}(k,x)h_{ij}(k,x)\|). \tag{3.53}$$

Now, a suitable conical bound on the interconnection terms would be as follows:

$$\sum_{j=1}^{N} \delta_i(\|e_{ij}(k,x)h_{ij}(k,x)\|) \leq \sum_{j=1}^{N} \bar{e}_{ij}\gamma_{ij}\delta_i(\|x_j\|), \tag{3.54}$$

for some nonnegative sector parameters γ_{ij}'s. Furthermore, the bounding of the increments of the Lyapunov function can be derived in a bit different way as follows:

$$v(k+1, x(k+1)) - v(k, x(k)) = \sum_{i=1}^{N} d_i(v_i(k+1, x(k+1)) - v_i(k, x(k)))$$
$$\leq \sum_{i=1}^{N} d_i(v_i(k+1, g_i(k, x_i(k), u_i) + h_i(k, x(k)))$$
$$- v_i(k, x(k)))$$
$$\leq \sum_{i=1}^{N} d_i(\Delta_g v(k, x(k)) + v_i(k+1, h_i(k, x(k))))$$
$$\leq \sum_{i=1}^{N} d_i(-l_i\delta_i(\|x_i(k)\|) + \sum_{j=1}^{N} \bar{e}_{ij}\gamma_{ij}\delta(\|x_j(k)\|))$$
$$= -d^T Q(\bar{E})\delta(\|x(k)\|) \tag{3.55}$$

where

$$\delta_i(\|x_j\|) = [\delta_1(\|x_1(k)\|, \ldots, \delta_N(\|x_N(k)\|)]^T. \tag{3.56}$$

As common by now, matrix

$$Q(\bar{E}) = [q_{ij}(\bar{E})] \tag{3.57}$$

is an $N \times N$ matrix with elements

$$q_{ij}(\bar{E}) = \begin{cases} l_i - \bar{e}_{ii}\gamma_{ii}, & i = j \\ -\bar{e}_{ij}\gamma_{ij}, & i \neq j \end{cases} \tag{3.58}$$

At this point, let us recall that $\bar{E}$ belongs to a convex hull with ℓ vertex matrices $\bar{E}^i$. Interesting to note from the technical point of view, even if we assume or check and it turns out to be true that all vertex matrices are M-matrices, every element of this polytope may not be an M-matrix. For a convex combination of two M-matrices there is a necessary and sufficient condition so that the whole polytope contains only M-matrices [12], stated in the following theorem, and proved in [25].

Theorem 3.3 *Let $\bar{E}^1$ and $\bar{E}^2$ be two M-matrices of the same dimension. Then, the following convex hull, that is a polytope,*

$$\bar{E}(\alpha) = \alpha\bar{E}^1 + (1-\alpha)\bar{E}^2, \ \alpha \in [0, 1], \tag{3.59}$$

contains elements so that each element is an M-matrix for the corresponding value of alpha if and only if matrix $(\bar{E}^2)^{-1}\bar{E}^1$ has no negative real eigenvalues.

Unfortunately, for a general number of M-matrix vertices there are only sufficient conditions [25]. On the other hand, the necessary and sufficient result for the case of two vertex matrices in Theorem 3.3 can be used to formulate other sufficient conditions for an arbitrary number of vertex M-matrices. This has been provided as Theorem 4.3 in [25], and reformulated as follows.

Theorem 3.4 *Let all ℓ vertex matrices $\bar{E}^i$ of the convex hull $\mathbb{E} = \text{conv}\{\bar{E}^i\}$ be M-matrices. Then, the convex hull contains only M-matrices if*

$$(\bar{E}^j)^{-1}\bar{E}^i$$

are M-matrices for all $j < i$, and $i, j \in \{1, \ldots, \ell\}$.

An important corollary of Theorem 3.4 is that it essentially guarantees that there is a common element by element positive vector d such that $\bar{E}^i d$ is also a positive vector element by element for each vertex matrix, that is, each $i \in \{1, \ldots, \ell\}$. The same property can be achieved if there is one vertex matrix that serves as a generator for the convex polytope in a way that is provided in Theorem 44 in [25] and reformulated in the following theorem.

Theorem 3.5 *Let all ℓ vertex matrices $\bar{E}^i$ of the convex hull $\mathbb{E} = \text{conv}\{\bar{E}^i\}$ be M-matrices. Then, all elements of the convex hull are M-matrices if there exits a generator vertex matrix $\bar{E}^l$ such that $(\bar{E}^l)^{-1}\bar{E}^i$ are M-matrices for all $i \neq l$, and $i \in \{1, \ldots, \ell\}$.*

The result, formulated in Theorem 3.5, also implies that there is a common element by element positive vector d such that $\bar{E}^i d$ is another positive element by element vector for each vertex matrix $i \in \{1, \ldots, \ell\}$. Therefore, the proofs of Theorems 3.4 and 3.5 are simple, and based on the existence of a common element by element positive vector d for all vertices. This implies, that the same vector would be suitable for any convex combination of the vertices, which establishes the proofs for both theorems in a simple manner.

Now, computation of an inverse M-matrix is not computationally efficient, in general [7, 8]. Therefore, a much more computationally efficient condition for guaranteeing the existence of a common positive vector d so that $\bar{E}^i d$ is an element by element positive vector for all vertex matrices, was formulated using linear programming in Theorem 4.4 in [25] and in Theorem 4.2 in [28]. It is used to establish connective stability for the structured interconnected system in the following theorem.

Theorem 3.6 *Let all of the ℓ vertex matrices $\bar{E}^i$ of the convex hull $\mathbb{E} = \text{conv}\{\bar{E}^i\}$ be M-matrices. Define the augmented matrices $\tilde{E}$ as $N \times (N+1)$ matrices such that*

$$\tilde{E}^i = [\bar{E}^i, \mathbf{1_N}]$$

where $\mathbf{1_N}$ is an N dimensional column vector with all elements equal to one. Define also an $(N+1)$ dimensional variable vector $z = [z_1, \ldots, z_{N+1}]^T$ and consider the following optimization problem:

$$\begin{aligned} &\max && z_{N+1} \\ &\textit{subject to} && \tilde{E}^i z \geq 0, \ \forall i \in \{1, \ldots, \ell\}, \\ & && z \geq 0, \ z_{N+1} \leq 1. \end{aligned} \tag{3.60}$$

Then, the structured discrete-time interconnected dynamical system is connectively stable if the result of the optimization is

$$z^o_{N+1} = \max z_{N+1} = 1.$$

Proof If the result of the optimization problem (3.60) turns out to be $z^o_{N+1} = \max z_{N+1} = 1$, then the convex hull contains only M-matrices and every $\bar{E}$ is an M-matrix, by Theorem 4.4 in [25] and Theorem 4.2 in [28]. Therefore, for any $\bar{E}$, there are two N dimensional and element by element positive vectors $d(\bar{E})$ and $c(\bar{E})$ such that

$$c^T(\bar{E}) = d^T(\bar{E})Q(\bar{E}).$$

This implies that for each $\bar{E}$ in the convex hull, there is a composite positive definite Lyapunov function that establishes stability of the corresponding structured discrete-time interconnected dynamical system, in the sense of Lyapunov, and thus the proof. □

If the vertex matrices are all M-matrices, then there is another linear programming optimization condition, which guarantees an existence of a common element by element positive vector d such that $(\bar{E}^i)^T d$ is an element by element positive vector for all vertex matrices. This result was provided in Theorem 4.3 in [28] and is used to formulate another, easy to compute, sufficient condition for connective stability of the structured discrete-time interconnected dynamical system in the following theorem.

Theorem 3.7 *Let all ℓ vertex matrices $\bar{E}^i$ of the convex hull $\mathbb{E} = \text{conv}\{\bar{E}^i\}$ be M-matrices. Define the augmented matrices $\hat{E}$ as $N \times (N+1)$ matrices such that $\hat{E}^i = [(\bar{E}^i)^T, \mathbf{1_N}]$ where $\mathbf{1_N}$ is an N dimensional column vector with all elements equal to one. Define also an $(N+1)$ dimensional variable vector $z = [z_1, \ldots, z_{N+1}]^T$ and consider the following optimization problem:*

$$\begin{aligned} \max \quad & z_{N+1} \\ \textit{subject to } & \hat{E}^i z \geq 0, \ \forall i \in \{1, \ldots, \ell\}, \\ & z \geq 0, \ z_{N+1} \leq 1. \end{aligned} \tag{3.61}$$

Then, the structured discrete-time interconnected dynamical system is connectively stable if the result of the optimization is

$$z^o_{N+1} = \max z_{N+1} = 1.$$

Proof If the outcome of the optimization problem (3.61) ends up being $z^o_{N+1} = \max z_{N+1} = 1$ then the whole convex hull contains only M-matrices so any fundamental interconnection matrix $\bar{E}$ is an M-matrix and thus the proof. □

If the vertex matrices are all M-matrices, whether the whole polytope contains only M-matrices as its elements, may also be checked using linear matrix inequalities as shown in Theorem 4.6 in [25] and Theorem 4.4 in [28]. This leads to another sufficient condition formulation of connective stability for the structured discrete-time interconnected dynamical system, which is computationally efficient to check. It is formulated in the following theorem.

Theorem 3.8 *Let all of the ℓ vertex matrices $\bar{E}^i$ of the convex hull $\mathbb{E} = \text{conv}\{\bar{E}^i\}$ be M-matrices. Then, consider the following linear matrix inequality feasibility problem:*

$$(\bar{E}^i)^T D + D\bar{E}^i \succ 0, \; \forall i \in \{1, \ldots, \ell\},$$
$$D \succ 0, \tag{3.62}$$

where D is a diagonal matrix. Then, the structured discrete-time interconnected dynamical system is connectively stable if the linear matrix inequality problem formulation is feasible.

Proof If the linear matrix inequality formulation (3.62) is feasible, then every $\bar{E}$ matrix will have the same diagonal matrix D establishing it is positive stable, which means it is an M-matrix and therefore the proof is completed. □

To conclude the presentation on connective stability, let us recapitulate what has been presented, that is, we formulated three sufficient conditions for connective stability of the structured discrete-time interconnected dynamical system based on convex programming. Two are based on linear programming and one on linear matrix inequalities convex optimization. In general, it is interesting to mention that there is a connection between these three conditions. This is formulated in Theorem 2.4 in [25] and reformulated in the following theorem.

Theorem 3.9 *Let E be an M matrix of dimension $N \times N$. If there exist two element by element positive vectors $d = [d_1, \ldots, d_N]^T$ and $\bar{d} = [\bar{d}_1, \ldots, \bar{d}_N]^T$ such that Qd and $Q^T\bar{d}$ are also element by element positive, then a positive definite diagonal matrix D constructed from d and $\bar{d}$ as $D = \text{diag}\{\bar{\mathrm{d}}_1/\mathrm{d}_1, \ldots, \bar{\mathrm{d}}_\mathrm{N}/\mathrm{d}_\mathrm{N}\}$ satisfies the following:*

$$E^T D + DE \succ 0.$$

This means that if the two connective stability formulations in Theorems 3.6 and 3.7 based on linear programming are both feasible, then the third one based on the linear matrix inequality convex programming formulated in Theorem 3.8 will be feasible too.

3.2 Computation of Decentralized Control Laws

So far we haven't provided any information on the actual computation of decentralized control laws for the overall discrete-time interconnected dynamical system. We only assumed their existence and that they stabilize the corresponding individual subsystem dynamics with estimated sector stability bounds evaluated in terms of positive definite functions $w_i(\cdot, \cdot)$, as formulated in inequalities (3.37). Regarding the actual computation of the decentralized control laws, we are restricted to only a few of the existing tools and algorithms. Again, one of the most efficient and convenient formulations with corresponding solutions if feasible, is based on the linear matrix inequality formulation and its convex programming way of solving. But then, similar to what was mentioned with respect to the unstructured case, this approach can be only applied to a very specific set of discrete-time structured dynamical systems. Therefore, let us assume that the subsystem discrete-time dynamics are of the following form:

$$x_i(k+1) = A_i x_i(k) + B_i u_i(k) + m_i(k, x(k)), i \in \{1, \ldots, N\}. \tag{3.63}$$

Given that the formulation is intended to lead to a linear matrix inequality convex optimization formulation, the nonlinear interconnection terms are assumed to satisfy the following conical constraints:

$$m_i^T(k, x) m_i(k, x) \leq \alpha_i^2 x^T M_i^T M_i x, i \in \{1, \ldots, N\}. \tag{3.64}$$

Here we use standard notation, that is, in Eq. (3.63), $x_i \in \mathbb{R}^{n_i}$ is the i-th subsystem's state vector, $u_i \in \mathbb{R}^{m_i}$ is its input vector and $m_i(k, x)$ is the subsystem's sector bounded perturbation. If we assume that each subsystems imposes a sector constraint (3.64) then the overall interconnected system will have N sector bounded constraints. In the case of multiple constraints, the *S-procedure* provides only sufficient conditions [4, 29] and furthermore a parameter that has to be treated as a variable accompanies each inequality and treating multiplicity of these parameters poses difficulties in formulating the inequalities as linear matrix inequalities. To bypass these difficulties all together, we consider a slightly different bounding of the interconnections with one single sector bounded condition in a compact form as follows:

$$m^T(k, x) m(k, x) \leq x^T \left(\sum_{i=1}^{N} \alpha_i^2 M_i^T M_i \right) x \tag{3.65}$$

where $m(k, x) = [m_1(k, x)^T, \ldots, m_N(k, x)^T]^T$, $A_D = \text{diag}\{A_1, A_2, \ldots, A_N\}$, and $B_D = \text{diag}\{B_1, B_2, \ldots, B_N\}$.

Now, we can formulate an overall convex programming problem in terms of linear matrix inequalities to be solved and if feasible provide a structure decentralized gain control law in the following form:

$$\begin{aligned}
&\min \left(\sum_{i=1}^{N} \gamma_i \right) \\
&\text{subject to} \\
&Q_D \succ 0 \\
&\gamma_i - \frac{1}{\bar{\alpha}_i^2} < 0, \;\; i \in \{1, 2, \ldots, N\} \\
&\begin{bmatrix}
-Q_D & Q_D A_D^T + L_D^T B_D^T & Q_D M_1^T & \ldots & Q_D M_N^T \\
A_D Q_D + B_D L_D & I - Q_D & 0 & \ldots & 0 \\
M_1 Q_D & 0 & -\gamma_1 I & \ldots & 0 \\
\vdots & \vdots & \vdots & \ddots & \vdots \\
M_N Q_D & 0 & 0 & \ldots & -\gamma_N I
\end{bmatrix} \prec 0
\end{aligned} \tag{3.66}$$

where

$$\alpha_i = 1/\sqrt{\gamma_i}, \; i \in \{1, 2, \ldots, N\}$$

and where $\bar{\alpha}_i$ denotes a minimal sector size for the i-th subsystem. To guarantee that the computed control law will be decentralized, if of course the problem is feasible, we assume that matrices Q_D and L_D are block diagonal, that is,

$$Q_D = \text{diag}\{Q_1, Q_2, \ldots, Q_N\}, \; L_D = \text{diag}\{L_1, L_2, \ldots, L_N\},$$

where the blocks match the subsystems' dimensions. Then, if the optimization problem (3.66) is feasible, the resulting overall decentralized control law can be computed as follows:

$$K_D = L_D Q_D^{-1}. \tag{3.67}$$

By construction, this control law is guaranteed to be in a block diagonal form, that is, decentralized. This is a consequence of the assumption that both matrices L_D and Q_D are block diagonal matrices, with the corresponding individual blocks of matching dimensions. Notice that this is only a sufficient condition for the gain matrix to be block diagonal, since there may be not necessarily block diagonal matrices Q and L that will result in LQ^{-1} being block diagonal. Unfortunately the parametrization of the gain matrix in terms of matrices L and Q in the optimization formulation (3.66), does not allow us to impose block diagonal structure as a convex constraint directly on the structure of K. Thus, our assumption provides a sufficient condition to guarantee the block diagonal structure of the feedback gain matrix K.

Among different structures, decentralized structure, for example, block diagonal in the linear case, is one of the simplest structures yet even in this case we face difficulties which cannot be handled in the necessary and sufficient way. More general and therefore more complex structures are even more difficult to integrate into the control design and thus imply even more computational difficulties. The incorporation of the structure into the control design is still very much an open problem, which is structure specific and its success intrinsically relies on the inherent structure of the system to be controlled. Naturally this imposes that the control law structure has to match the corresponding system's structure. Another avenue worth exploring is to consider if general decompositions of functions [1, 16] can be utilized for structural decompositions of dynamical systems being represented by either differential or difference vector equations where their righthand sides are being decomposed as functions of multiple state variables.

References

1. V. I. Arnoíd. On functions of three variables. *Doklady Akademii Nauk SSSR*, vol. 114, pp. 953–956, 1957.
2. A. Berman and R. J. Plemmons. *Nonnegative Matrices in Mathematical Sciences*. Academic Press, New York, NY, 1979.
3. R. Bellman. Vector Lyapunov functions. *SIAM Journal of Control*, vol. 1, pp. 32–34, 1962.

4. S. Boyd, L. E. Ghaoui, E. Feron, and V. Balakrishnan, *Linear Matrix Inequalities in System and Control Theory*, Philadelphia, PA, SIAM, 1994.
5. S. Boyd and L. Vandenberghe. *Convex Optimization*. Cambridge University Press, 2004.
6. P. Gahinet, A. Nemirovski, A. J. Laub, and M. Chilali. *LMI Control Toolbox: For Use with MATLAB*. The MathWorks, Natick, MA, 1995.
7. F. R. Gantmacher. *The Theory of Matrices*. AMS Chelsea Publishing, New York, NY, 1960.
8. G. H. Golub and C. F. Van Loan. *Matrix Computations*. 4th Edition, Johns Hopkins University Press, Philadelphia, PA, 2013.
9. M. W. Hirsch, S. Smale, and R. L. Devaney, *Differential Equations, Dynamical Systems, and an Introduction to Chaos*, 3rd ed., Waltham, MA: Elsevier Inc., 2013.
10. D. W. C. Ho and G. Lu. Robust stabilization for a class of discrete-time non-linear systems via output feedback: the unified LMI approach. *International Journal of Control*, vol. 76, no. 2, pp. 105–115, 2003.
11. R. A. Horn and C. R. Johnson. *Matrix Analysis*. Cambridge University Press, 1985.
12. R. A. Horn and C. R. Johnson. *Topics in Matrix Analysis*. Cambridge University Press, 1991.
13. R. E. Kalman and J. E. Bertram. Control system analysis and design via the "Second method of Lyapunov": I Continuous-time systems. *Journal of Basic Engineering, Transactions of the ASME*, vol. 82, pp. 371–393, 1960.
14. R. E. Kalman and J. E. Bertram. Control system analysis and design via the "Second method of Lyapunov": II Discrete-time systems. *Journal of Basic Engineering, Transactions of the ASME*, vol. 82, pp. 394–400, 1960.
15. H. K. Khalil. *Nonlinear Systems*, 3rd Edition., Upper Saddle River, NJ: Prentice Hall, 2002.
16. N. N. Kolmogorov. On the representation of continuous functions of many variables by superposition of continuous functions of one variable and addition. *Doklady Akademii Nauk SSSR*, 144:679–61, 1957; *American Mathematical Society Translation*, vol. 28, pp. 55–59, 1963.
17. J. P. LaSalle, *The Stability and Control of Discrete Processes*. New York, NY, Springer-Verlag, 1986.
18. V. M. Matrosov. On the theory of stability of motion. *Prikladnia Matematika i Mekhanika*, vol. 26, pp. 992–1000, 1962.
19. K. M. Miettinen. *Nonlinear Multiobjective Optimization*. Kluwer Academic Publishers, Norwell, MA, 1998.
20. D. A. Specher. On the Structure of Continuous Functions of Several Variables. *Transactions of the American Mathematical Society*, vol. 115, pp. 340–355, 1965.
21. A. W. Starr and Y. C. Ho. Nonzero-sum differential games. *Journal of Optimization Theory and Applications*, vol. 3, pp. 184–206, 1969.
22. A. W. Starr and Y. C. Ho. Further properties of nonzero-sum differential games. *Journal of Optimization Theory and Applications*, vol. 3, pp. 207–219, 1969.
23. D. D. Šiljak. *Large-scale Dynamic Systems: Stability and Structure*. North-Holland, New York, NY, 1978.
24. D. D. Šiljak. *Decentralized Control of Complex Systems*. Academic Press, Boston, MA, 1991.
25. D. M. Stipanović and D. D. Šiljak. Stability of polytopic systems via convex M-matrices and parameter-dependent Liapunov functions. *Nonlinear Analysis*, vol. 40, pp. 589–609, 2000.
26. D. M. Stipanović and D. D. Šiljak. Jacobi and Gauss-Seidel Iterations for Polytopic Systems: Convergence via convex M-matrices. *Reliable Computing*, vol. 6, pp. 123–137, 2000.
27. D. M. Stipanović and D. D. Šiljak. Robust stability and stabilization of discrete-time non-linear systems: The LMI approach. *International Journal of Control*, vol. 74, pp. 873–879, 2001.

28. D. M. Stipanović and D. D. Šiljak. Connective stability of discontinuous dynamic systems. *Journal of Optimization Theory and Applications*, vol. 115, pp. 711–726, 2002.
29. V. A. Yakubovich. The S-procedure in nonlinear control theory. *Vestnik Leningradskogo Universiteta*, vol. 1, pp. 62–77, 1971. In Russian, English translation is in Vestnik Leningrad. Univ. Math, vol. 4, pp. 73-93.

Part II
Neural Networks and Machine Learning

4 Stability of Difference Equations and Long Short-Term Memory Neural Networks

By now it has been a while that long short-term memory neural networks [6, 9] have been successfully used to process time-varying sequences [5]. While they have been very useful in implementations their properties as discrete-time dynamical systems expressed in terms of difference equations are still quite a bit of a mystery. One may stipulate that this is, at least partially, due to the use of specific sigmoid functions and their concatenations with multiple weight matrices and biases, as they appear in the long short-term memory neural network dynamical model. This model carries some inherent difficulties where even most basic questions like the existence and the number of equilibria, are complex to answer due to the transcendental nature of the corresponding fixed-point algebraic equation, which they are solutions to. Therefore, the heuristic implementations of the long short-term memory neural networks have been successful yet there is a lack of technical analysis for their complex dynamic behaviors. In this chapter we recall the analysis provided in [15] and we continue in the same direction of finding connections between the nonlinear autonomous long short-term memory neural network dynamical model and its linearization. More sophisticated simulations are provided to support the idea that these two discrete-time dynamical systems behave similarly in terms of the equilibria stability, yet unlike in [15], we were able to determine some discrepancies, at least in simulations.

The autonomous long short-term memory neural network dynamical model is chosen because of its intriguing behaviors from the difference equations point of view and the link between its chaotic behavior and proper training in the machine learning applications. As it will be illustrated in this chapter, long short-term memory neural networks have quite profound dynamic properties and produce very interesting chaotic behaviors [7, 8]. As autonomous dynamical systems, that is, when the input or the reference signal is not present, they are attractive with respect to the unit box since all important, interesting and complex dynamic behaviors happen inside the box. One of the long recognized characteristics of the

D. Stipanović, *Difference Equations and Machine Learning*, Synthesis Lectures on Mathematics & Statistics, https://doi.org/10.1007/978-3-032-00910-4_4

trained recurrent neural networks (which include long short-term memory neural networks) is that they, if trained to perform their tasks well, do operate at the so called "edge of chaos." Also, the edge of chaos can be equivalently considered as the edge of stability [1] given that the edge, in this case, seems to be the boundary between the chaotic and the stability regions. Furthermore, this edge or boundary is also a link between the chaotic behavior and the machine learning application performance. It turns out, at least through the simulations of the lower order autonomous long short-term memory neural networks, that the boundary of stability of the linearized model is a surprisingly good characterization of the "edge of chaos." Due to being quite unusual, the chaos or the chaotic behaviors are behaviors that cannot be classified as any other standard behaviors [7, 8]. Again, the importance of the"edge of chaos" stems from the so far never contradicted observation that it is connected to the recurrent neural networks' learning efficiency [2].

4.1 Long Short-Term Memory Neural Network Discrete-Time Model Linearization

The material presented in this section is dealing with the specific and quite close relationship between behaviors of the linear and nonlinear models of the autonomous long short-term memory neural networks [15]. As a dynamical system, the nonlinear long short-term memory dynamical model is nothing but a multidimensional difference equation. Without the input or reference variable, the nonlinear model simply can be represented as follows [5]:

$$\begin{aligned} h(k+1) &= \varphi(W_o h(k) + b_o) \odot \tanh(\varphi(W_f h(k) + b_f) \\ &\quad \odot c(k) + \varphi(W_i h(k) + b_i) \odot \tanh(W_g h(k))), \\ c(k+1) &= \varphi(W_f h(k) + b_f) \odot c(k) \\ &\quad + \varphi(W_i h(k) + b_i) \odot \tanh(W_g h(k)), \end{aligned} \tag{4.1}$$

where k is a nonnegative integer representing discrete time,

$$h(\cdot) = [h_1(\cdot), \ldots, h_n(\cdot)]^T$$

is an n-dimensional output state variable vector, and

$$c(\cdot) = [c_1(\cdot), \ldots, c_n(\cdot)]^T$$

is an n-dimensional memory state variable vector. Function

$$\varphi(\cdot) = \frac{e^{(\cdot)}}{1 + e^{(\cdot)}} \tag{4.2}$$

is the standard logistic function and $\tanh(\cdot)$ is the hyperbolic tangent function that are applied element-wise, and $\odot$ represents the Hadamard product. W_o, W_f, W_i, and W_g are $n \times n$

weight matrices assumed almost exclusively in practical implementations to be constant and b_o, b_f, and b_i are n-dimensional bias vectors, also most commonly assumed to be constant. These assumptions of both weight matrices and bias vectors to be constant are interesting to discuss from the control design point of view. The reference input signal is not a control variable yet it can be considered as a desired trajectory to be followed by the output of the network. The weight matrices and bias vectors are computed during the training, and thus they are essentially the control variables. So for the most of the machine learning applications, the control design is restricted to computing these variables during the training, as constants. For the purposes of analyzing and presenting dynamic behaviors of the long short-term memory neural networks in a neater way, and following [15] the gate bias vector b_g is set to zero, so that the origin is an equilibrium of the overall discrete-time system (4.1). This way, a numerically complex computation of the nonzero equilibria is bypassed, at least for now.

In terms of the nonlinear model's general behavior characteristics, Eq. (4.1) produce

$$h(k+1) = \varphi(W_o h(k) + b_o) \odot \tanh(c(k+1)), \tag{4.3}$$

which implies

$$|h_j(k)| \leq |c_j(k)| \text{ and } \operatorname{sign}(h_j(k)) = \operatorname{sign}(c_j(k)),\ \forall j \in \{1, \ldots, n\}, \text{ and } \forall k. \tag{4.4}$$

This simple observation, stating that the norm of the output state variable is bounded by the norm of the memory state variable, means that the stability properties of the state variables are related. Technically speaking, variable $c(k)$ can be expressed in terms of $h(k+1)$ and $h(k)$ variables from the first vector difference equation in (4.1) and then substituted in the second vector difference equation. That would result in a second order implicit difference equation in terms of $h(\cdot)$ only, yet the form would be more cumbersome to analyze.

The local dynamic behavior analysis of the nonlinear model around its equilibrium at the origin can be carried out in a standard way, that is by computing the Jacobian for an equilibrium at zero and studying its properties dependent on the nonlinear model parameters. The parameters are entries of the weight matrices and bias vectors. Because after the training they are all set to constant values, it is appropriate to treat them as parameters, at least for the purposes of the stability analysis. In order to compute the Jacobian we can streamline the presentation by denoting the derivative with respect to an independent variable as D. This notation will be used only in this chapter and for the derivation that follows. Using this notation, the derivatives of the logistic and hyperbolic functions are simply computed as follows:

$$D\varphi(\cdot) = \varphi(\cdot)(1 - \varphi(\cdot)),\ D\tanh(\cdot) = 1 - \tanh^2(\cdot). \tag{4.5}$$

Following the line of thought presented in [15], we will illustrate first how to compute the Jacobian when $n = 2$ and then show how this result generalizes to an arbitrary dimension n. In the case when $n = 2$ the state variable vector is of dimension four and we can write four righthand side components of Eq. (4.1) as follows:

$$
\begin{aligned}
F_1(h,c) = &\ \varphi(W_o^1 h + b_o^1)\tanh(\varphi(W_f^1 h + b_f^1)c_1 \\
&+\varphi(W_i^1 h + b_i^1)\tanh(W_g^1 h)), \\
F_2(h,c) = &\ \varphi(W_o^2 h + b_o^2)\tanh(\varphi(W_f^2 h + b_f^2)c_2 \\
&+\varphi(W_i^2 h + b_i^2)\tanh(W_g^2 h)), \\
F_3(h,c) = &\ \varphi(W_f^1 h + b_f^1)c_1 + \varphi(W_i^1 h + b_i^1)\tanh(W_g^1 h), \\
F_4(h,c) = &\ \varphi(W_f^2 h + b_f^2)c_2 \\
&+\varphi(W_i^2 h + b_i^2)\tanh(W_g^2 h),
\end{aligned} \tag{4.6}
$$

where superscripts denote rows. To linearize the system of nonlinear difference equations in Eq. (4.1) we proceed by computing Jacobian matrix $L = [l_{ij}]$ through computing corresponding partial derivatives and evaluating them at zero, which is due to the equilibrium being at the origin. Therefore,

$$
\begin{aligned}
l_{11} = &\ \frac{\partial F_1(h,c)}{\partial h_1}\bigg|_{h=c=0} = D\varphi(b_o^1)W_o^{11}\tanh(0) \\
&+\varphi(b_o^1)D\tanh(0)(D\varphi(b_f^1)W_f^{11}\ c_1|_{c1=0} \\
&+D\varphi(b_i^1)W_i^{11}\tanh(0) + \varphi(b_i^1)D\tanh(0)W_g^{11}) \\
= &\ \varphi(b_o^1)\varphi(b_i^1)W_g^{11}.
\end{aligned} \tag{4.7}
$$

Following the same derivation technique we can derive the the following coefficients:

$$
\begin{aligned}
l_{12} &= \frac{\partial F_1(h,c)}{\partial h_2}\bigg|_{h=c=0} = \varphi(b_o^1)\varphi(b_i^1)W_g^{12}, \\
l_{21} &= \frac{\partial F_2(h,c)}{\partial h_1}\bigg|_{h=c=0} = \varphi(b_o^2)\varphi(b_i^2)W_g^{21}, \\
l_{22} &= \frac{\partial F_2(h,c)}{\partial h_2}\bigg|_{h=c=0} = \varphi(b_o^2)\varphi(b_i^2)W_g^{22}.
\end{aligned} \tag{4.8}
$$

Practically repeating the procedure and performing the same derivation steps lead to the following:

$$
\begin{aligned}
l_{13} &= \frac{\partial F_1(h,c)}{\partial c_1}\bigg|_{h=c=0} = \varphi(b_o^1)D\tanh(0)\varphi(b_f^1) \\
&= \varphi(b_o^1)\varphi(b_f^1), \\
l_{14} &= \frac{\partial F_1(h,c)}{\partial c_2}\bigg|_{h=c=0} = 0,
\end{aligned} \tag{4.9}
$$

and furthermore

$$l_{23} = \left.\frac{\partial F_2(h,c)}{\partial c_1}\right|_{h=c=0} = 0,$$
$$l_{24} = \left.\frac{\partial F_2(h,c)}{\partial c_2}\right|_{h=c=0} = \varphi(b_o^2)\varphi(b_f^2). \tag{4.10}$$

When computing the remaining two rows of the Jacobian matrix L, the only difference is that the elements do not depend on the bias vector b_o. By introducing diagonal matrices

$$B_j = \text{diag}\{\varphi(b_j^1), \varphi(b_j^2)\}, \ \forall j \in \{o, i, f\}, \tag{4.11}$$

we can write the Jacobian in a compact form as follows:

$$L = \begin{bmatrix} B_o B_i W_g & B_o B_f \\ B_i W_g & B_f \end{bmatrix}. \tag{4.12}$$

This expression in Eq. (4.12), is general in nature, and without simple and unnecessary technical arguments that will just clutter the presentation, it can be shown it is valid for the general nonlinear $2n$ dimensional autonomous long short-term memory neural network model. This Jacobian can be further simplified and used to establish local, near the origin, behavior characteristics, as it will be shown in what follows.

Now, we can formulate a local Lyapunov asymptotic stability result, in the sense of Lyapunov [11, 12], for the $2n$ dimensional nonlinear autonomous long short-term memory network model in the following theorem.

Theorem 4.1 *The nonlinear, time-invariant and autonomous $2n$ dimensional long short-term neural network dynamical model (4.1) is locally asymptotically stable with respect to the origin as an equilibrium, if all eigenvalues of an $n \times n$ matrix $B_i W_g B_o + B_f$ have absolute values less than one, that is, all lie strictly inside the unit circle.*

Proof The proof is based on some matrix transformations and manipulations. The first one is to note that Jacobian L in Eq. (4.12), can be represented as follows:

$$L = \begin{bmatrix} B_o & O_n \\ O_n & I_n \end{bmatrix} \begin{bmatrix} B_i W_g & I_n \\ B_i W_g & I_n \end{bmatrix} \begin{bmatrix} I_n & O_n \\ O_n & B_f \end{bmatrix} \tag{4.13}$$

where O_n and I_n stand for the $n \times n$-dimensional zero and identity matrices, respectively. It is a well known fact that for any two square matrices C and D the eigenvalues of CD are exactly the same as the eigenvalues of DC. Therefore, the eigenvalues of L are the same as the eigenvalues of $\tilde{L}$ where

$$\tilde{L} = \begin{bmatrix} B_i W_g & I_n \\ B_i W_g & I_n \end{bmatrix} \begin{bmatrix} B_o & O_n \\ O_n & B_f \end{bmatrix} = \begin{bmatrix} B_i W_g B_o & B_f \\ B_i W_g B_o & B_f \end{bmatrix}. \tag{4.14}$$

Another well known result is that the eigenvalues of a square matrix stay the same if the matrix is multiplied from one side by a nonsingular square matrix of the same dimension, and on the other side by its inverse. By choosing this nonsingular matrix to be

$$S = \begin{bmatrix} I_n & -I_n \\ O_n & I_n \end{bmatrix}, \tag{4.15}$$

implying its inverse can be computed as

$$S^{-1} = \begin{bmatrix} I_n & I_n \\ O_n & I_n \end{bmatrix} \tag{4.16}$$

we obtain that the following matrix:

$$\bar{L} = S\tilde{L}S^{-1} = \begin{bmatrix} O_n & O_n \\ B_i W_g B_o & B_i W_g B_o + B_f \end{bmatrix} \tag{4.17}$$

has the same eigenvalues as $\tilde{L}$, and thus the same eigenvalues as the system's Jacobian matrix L. Given that $\bar{L}$ in Eq. (4.17), is in a lower block triangular form, n of its eigenvalues are placed at zero, and thus stable, and the remaining n are the eigenvalues of $B_i W_g B_o + B_f$. Therefore, the overall system is locally asymptotically stable if all eigenvalues of $B_i W_g B_o + B_f$ have absolute values less than one. Thus, the proof is completed. □

As it will be shown in the simulation examples in the following section, the linear and nonlinear models have remarkably matching behaviors in terms of stability with respect to the equilibrium at the origin. It is expected that they should behave the same way locally, given the known stability properties of the linearization around the equilibrium, but what the simulations have shown is that this similarity goes beyond local. This is still an open and intriguing phenomenon for which there are no general technical explanations yet. But nothing stops us from hypothesizing what could be the potential roots of this phenomenon. One may be the properties of the sigmoid functions, like a simple property of the hyperbolic tangent function being

$$|\tanh(z)| \leq |z|$$

and of the logistic function (4.2) being bounded between zero and one. Thus, it looks like that the overall system is a good candidate to be considered as a concatenation of the linear and sector bounded nonlinear part that has been already treated in this book using linear matrix inequality optimization formulation [10, 14]. To streamline the presentation and make it compatible with the rest of the text we can introduce the overall state vector $x(\cdot) = [h(\cdot)^T, c(\cdot)^T]^T$ so that the system difference Eq. (4.1) can be rewritten in a more compact form as

$$
\begin{aligned}
x(k+1) = f(x(k)) &= Lx(k) + f(x(k)) - Lx(k) \\
&= Lx(k) + g(x(k)),
\end{aligned}
\tag{4.18}
$$

where the nonlinearity $g(\cdot)$ is simply defined by $g(x) \equiv f(x) - Lx$. Now, if W_g is chosen so that L is stable then the following linear matrix inequality optimization/minimization problem, in terms of scalar variable γ and symmetric matrix variable $P = P^T$, is known to be feasible [10, 14]:

$$
\begin{aligned}
&\min \gamma \\
&\text{subject to} \\
&P \succ 0, \\
&\begin{bmatrix} L^T P L - P & L^T P & G^T \\ PL & P - I & 0 \\ G & 0 & -\gamma I \end{bmatrix} \prec 0.
\end{aligned}
\tag{4.19}
$$

This implies that the nonlinear discrete-time dynamical system (4.18) is robustly stable for any nonlinear function $g(\cdot)$ satisfying the following sector bound:

$$
g^T(x)g(x) \leq \alpha^2 x^T G^T G x,\ \alpha = \frac{1}{\sqrt{\gamma}}. \tag{4.20}
$$

Positive definite matrix P defines a quadratic Lyapunov function $V(x(k)) = x(k)^T P x(k)$, which establishes this robust stability result. So, the stability of the nominal linear part guarantees the feasibility of the problem and the minimization of γ results in the maximization of the sector size parameter α.

How to estimate region of attraction based on the multi-variable mean-value theorem has been presented in [3]. In combination with the fact that the nonlinear autonomous long short-term memory discrete-time model is attractive and invariant with respect to the unit box, this derivation has been used to establish global asymptotic stability conditions and furthermore used to provide a computational method for weights and biases to stabilize the system. More computationally efficient global stabilization conditions using a linear matrix inequality convex optimization formulation have been presented in [4].

4.2 Autonomous Long Short-Term Memory Neural Network Dynamic Behavior Illustrations

As stated earlier in this chapter, there is still no precise technical characterization of the boundary of chaotic regime of operation for the autonomous long short-term memory neural networks. Therefore, in this section, we provide at least some representative simulations illustrating how interestingly accurate is the linear model characterization of the boundary of stability region of the corresponding nonlinear model. What makes this even more interesting is that the boundary of the stability region is also a boundary of the chaotic behavior region, that is, what we refer to as the "edge of chaos."

The first simulation is the first step in a build up toward the numerical characterization of the "edge of chaos." It is chosen for the purpose of rerunning a previously introduced example from [13] with the following weight matrices:

$$W_i = \begin{bmatrix} -1 & -4 \\ -3 & -2 \end{bmatrix}, \; W_o = \begin{bmatrix} 4 & 1 \\ -9 & -7 \end{bmatrix},$$
$$W_f = \begin{bmatrix} -2 & 6 \\ 0 & -6 \end{bmatrix}, \; W_g = \begin{bmatrix} -1 & -6 \\ 6 & -9 \end{bmatrix},$$

with all bias vectors being set to zero. So this example numerical values match our analysis for the behaviors of autonomous nonlinear long short-term memory neural network model with an equilibrium at the origin. This is basically implied by setting all bias vectors to zero, and thus the gate bias vector too. The simulation is initialized to start very closely to the origin (which is an equilibrium), with initial values chosen to be 10^{-50} for all four variables. This has been done since the linear model is unstable with two nonzero eigenvalues of L equal to $-0.7500 \pm 1.1180i$. The simulation is run for 10^5 time samples. For the purpose of a neater illustration, a plot of h_2 versus h_1, for the last 50 samples of simulation, is given in Fig. 4.1. As linear system model indicates, the nonlinear system is stable and it diverges from the origin. Therefore, the origin behaves as an unstable equilibrium and the outcome coincides with the statements and observations reported in [13].

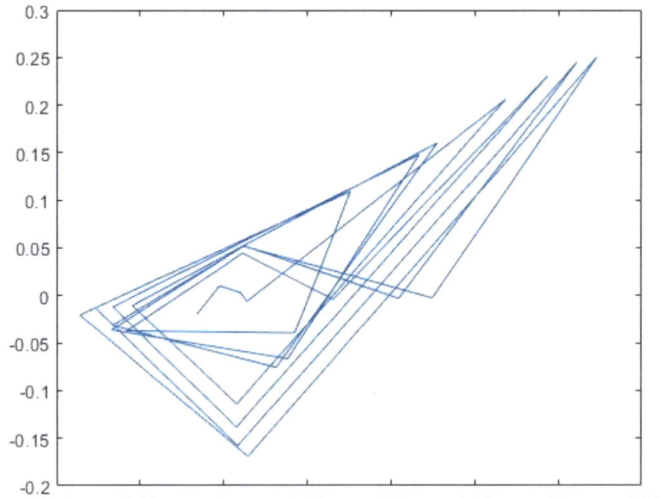

Fig. 4.1 h_2 versus h_1 when $W_g = [-1, -6; 6, -9]$

Next, while keeping the other matrices the same, we change numerical values for the entries of matrix W_g so that it is still unstable but with nonzero eigenvalues closer to the unit circle, which results in the following choice:

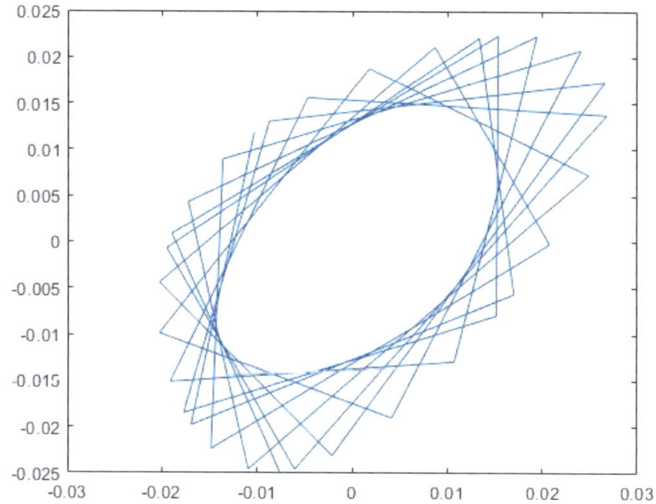

Fig. 4.2 h_2 versus h_1 when $W_g = [-0, 45, -4.55; 4.55, -5]$

$$W_g = \begin{bmatrix} -0.5 & -4.6 \\ 4.6 & -5 \end{bmatrix}.$$

In this case, the nonzero eigenvalues of L are $-0.1875 + \pm 1.0030$ with the absolute value of 1.0204. We use the same initial conditions at 10^{-50} for all variables, run the simulation for 10^5 samples and plot only the last 50 samples. Now, the plot of last 50 samples of h_2 versus h_1, is provided in Fig. 4.2. It can be observed from this figure that the autonomous nonlinear long short-term memory neural network model behaves in an unstable way, where the linear model is unstable with its eigenvalues being outside of the unit circle.

By changing numerical values for the entries of matrix W_g once again, while again keeping the values of other weight matrices the same, and the bias vectors still set to zero, to the following:

$$W_g = \begin{bmatrix} -0.45 & -4.55 \\ 4.55 & -5 \end{bmatrix},$$

the nonzero eigenvalues of L change to $-0.1813 \pm 0.9851i$ with the absolute value of 1.0016. Therefore, with this choice of numerical values the nonzero eigenvalues are even closer to the unit circle yet still outside of it. The simulation is now set to run for 10^6 samples and the last 50 samples of h_2 versus h_1 graph are depicted in Fig. 4.3. This indicates an unstable behavior since furthermore if the same simulation is run for 10^7 samples the same range of values for the state variables are obtained. Notice that the range of values of the state variables, for the last 50 samples, is being of order 10^{-3}. As observed, the nonlinear autonomous long short-term neural network model still exhibits an unstable behavior, which

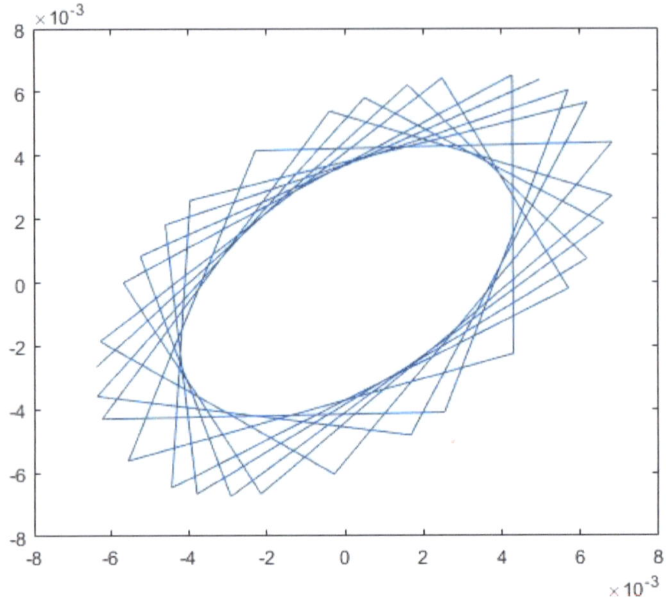

Fig. 4.3 h_2 versus h_1 when $Wg = [-0.5, -4.51; 4.51, -5]$

again matches the behavior of its linearized model, given that the nonzero eigenvalues lie outside the unit circle.

The next step is to place the eigenvalues of the linearized system inside the unit circle, that is, inside the stable area. Therefore, the goal is to choose numerical values of the entries of matrix W_g so that all eigenvalues of L are inside the unit circle. One such choice is when we set the following numerical values:

$$W_g = \begin{bmatrix} -0.45 & -4.52 \\ 4.55 & -5 \end{bmatrix}$$

in which case the nonzero eigenvalues of L are $-0.1812 \pm 0.9808i$ with the absolute value equal to 0.9974. Given that this is now the case of a stable linearized model, we choose the initial conditions to be equal to 10 for each one of the four variables. By doing so, the simulation starts outside of the unit box and thus far away from the equilibrium at the origin, and therefore this scenario becomes quite drastic and representative for a stable scenario. We run the simulation for 10^5 samples and the last 50 samples of h_2 versus h_1 are depicted in Fig. 4.4. In this case we also get a matching between nonlinear and linear models, in terms of their dynamic behaviors. Notice that the axes range values for the last 50 samples are of order 10^{-117}. If the simulation is run with the same initial conditions just twice as long, that is, for 2×10^5 samples, it can be observed from Fig. 4.5 that the order of magnitude for both variables drops to 10^{-231}. Thus, the trajectory clearly shows convergence toward

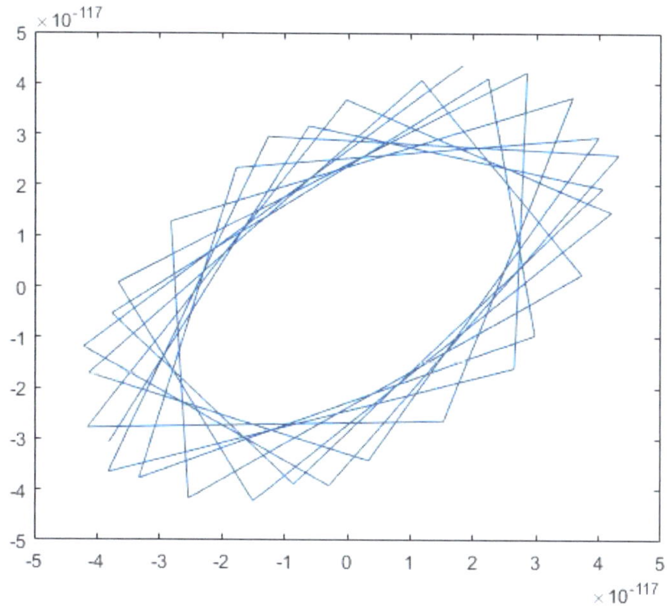

Fig. 4.4 h_2 versus h_1 when $Wg = [-0.45, -4.52; 4.55, -5]$ run for 10^5 samples

the origin, which indicates stable behavior of the autonomous nonlinear long short-term memory neural network model.

All simulations shown so far suggest that the local asymptotic stability, which has been derived to depend only on W_g (in terms of weight matrices), is a very good indicator of the autonomous nonlinear long short-term memory neural network stability behavior with respect to the origin. We also need to note that (c_1, c_2) graphs were plotted for each case and they appear very similar to (h_1, h_2) graphs (appearing as just being scaled) and thus have not been presented. The stability and instability behaviors between variables $c(\cdot)$ and $h(\cdot)$ have been observed as synchronized, as expected, in each case.

Similar behavior has been observed by increasing the system's dimension to six. This means that we increased dimensions of the state vectors h and c to three. For the ease of controlling the placement of the eigenvalues, all bias vectors were kept at zero. Furthermore, setting the gate bias vector entries to zero enables us to easily compute the Jacobian of the linearized model. As the first illustration, we consider the system with the following numerical values for the weight matrices:

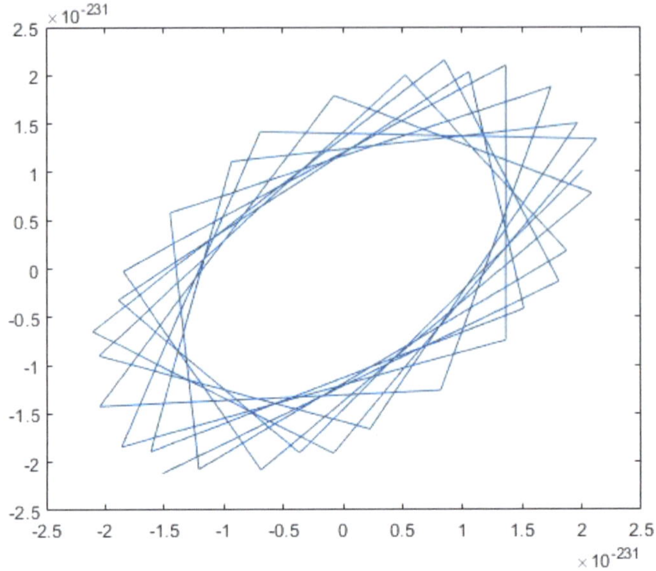

Fig. 4.5 h_2 versus h_1 when $Wg = [-0.45, -4.52; 4.55, -5]$ run for 2×10^5 samples

$$W_i = \begin{bmatrix} -1 & -4 & -2 \\ -3 & -2 & 2 \\ 1 & 2 & 3 \end{bmatrix}, \quad W_o = \begin{bmatrix} 4 & 1 & -1 \\ -9 & -7 & 1 \\ 2 & -3 & 4 \end{bmatrix}$$

$$W_f = \begin{bmatrix} -2 & 6 & -1 \\ 0 & -6 & 2 \\ 0 & 1 & 1 \end{bmatrix}, \quad W_g = \begin{bmatrix} -1 & -2.95 & 1 \\ 4.5 & 0 & 1 \\ -2.2 & 0 & -2 \end{bmatrix}.$$

In this case, the three nonzero eigenvalues are $0.2893 \pm 0.9536i$, and 0.1713 with absolute values equal to 0.9965 and 0.1713, respectively. Given the system is stable we set the initial conditions to 10 for all six variables. This means that the system trajectories start far away from the box where the equilibrium is at the origin, which is a representative scenario for the stable behavior. The simulation is run for 210^5 samples while only last 30 samples for the first two variables of h are depicted in Fig. 4.6.

If we now plot the third variable of h versus the first one, as depicted in Fig. 4.7, we can conclude that all three h variables, and thus all six variables, show convergence toward the origin.

Let us now slightly change the numerical values of the entries of the gate weight matrix as follows:

$$W_g = \begin{bmatrix} -1 & -3 & 1 \\ 4.5 & 0 & 1 \\ -2.2 & 0 & -2 \end{bmatrix},$$

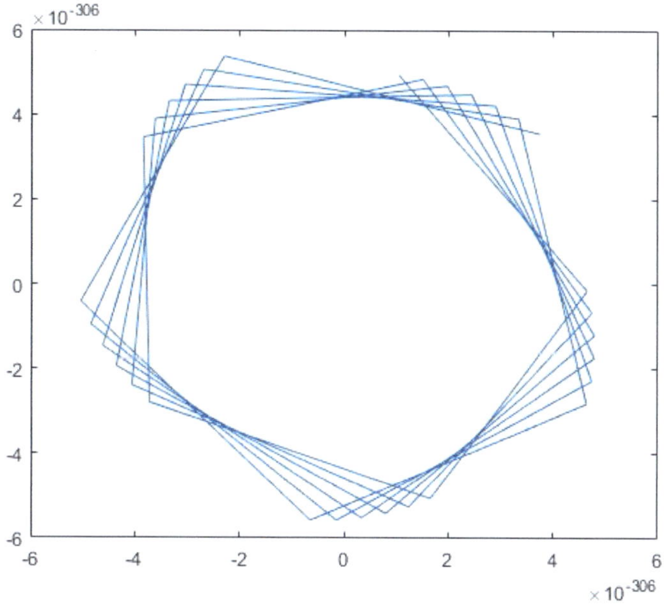

Fig. 4.6 Last 30 samples of h_2 versus h_1 for the stable six dimensional example run for 2×10^5 samples

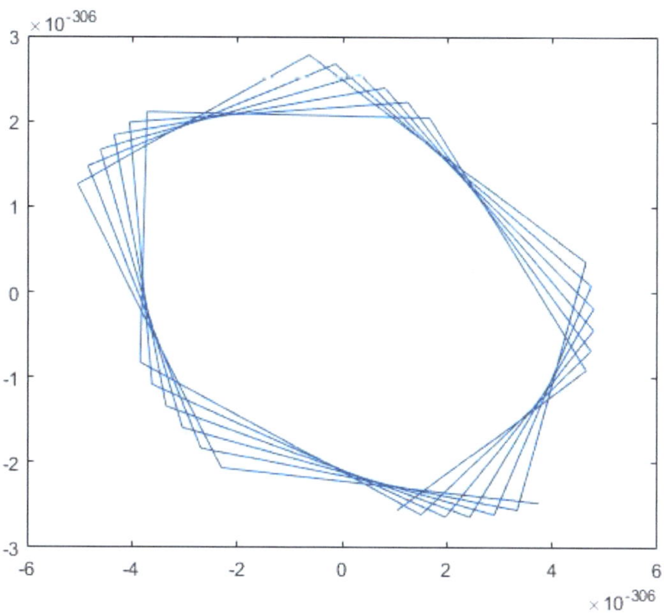

Fig. 4.7 Last 30 samples of h_3 versus h_1 for the stable six dimensional example run for 2×10^5 samples

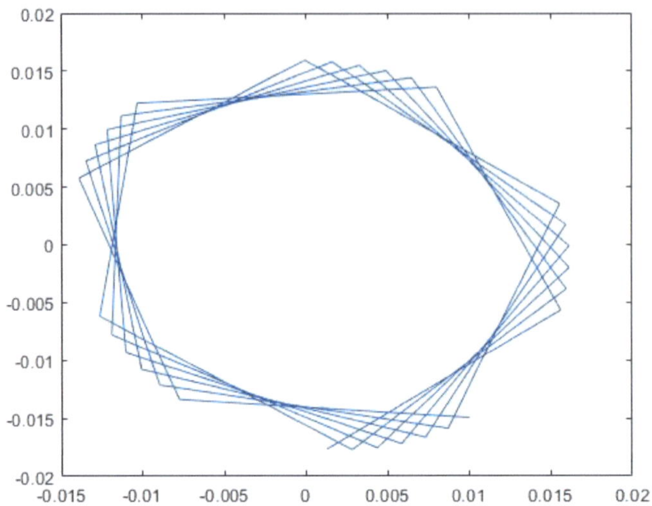

Fig. 4.8 Last 30 samples of h_2 versus h_1 for the unstable six dimensional example run for 2×10^5 samples

which results in nonzero eigenvalues being $0.2897 \pm 0.9610i$ and 0.1706 with absolute values of 1.0037 and 0.1706, respectively. Now given the linear model is unstable we impose the initial conditions to be close to the origin, that is, all variables start from 10^{-50}. Last 30 samples of the first two variables of h are depicted in Fig. 4.8.

To explore further the linear and nonlinear model matching dynamic behaviors, the degree of the discrete-time long short-term memory neural network model, is increased even more so that h vector variable has four scalar variables implying the same number of variables for c. This means that the dimension of the overall model is increased to eight. Again, we set all bias vectors to zero to have an easier control over the placement of the Jacobian eigenvalues and so that the origin is still an equilibrium. The numerical values for the entries of the weight matrices are chosen as follows:

$$W_i = \begin{bmatrix} -1 & -4 & -2 & 6 \\ -3 & -2 & 4 & 2 \\ 1 & 2 & 3 & -2 \\ 5 & -2 & 4 & 6 \end{bmatrix}, \quad W_o = \begin{bmatrix} 4 & 1 & -1 & 2 \\ -9 & -7 & 1 & 6 \\ 2 & -3 & 4 & 1 \\ -3 & 4 & -6 & 5 \end{bmatrix}$$

$$W_f = \begin{bmatrix} -2 & 6 & -1 & 9 \\ 0 & -6 & 2 & 3 \\ -2 & 4 & 5 & 1 \\ -7 & 3 & 1 & -1 \end{bmatrix}, \quad W_g = \begin{bmatrix} -1 & -3 & 2 & 0.5 \\ 3.9 & 0 & 1 & -0.1 \\ -2.2 & 0 & -2 & 0.5 \\ 1 & -1 & 0.5 & 1 \end{bmatrix}.$$

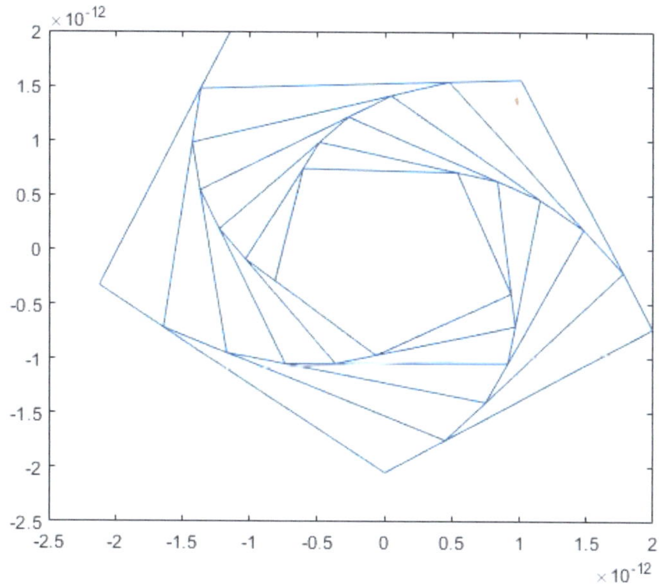

Fig. 4.9 Last 30 samples of h_2 versus h_1 for the eight dimensional example run for thousand samples starting from the initial conditions equal to 0.1

and result in nonzero eigenvalues of L located at $0.2535 \pm 0.9360i$, 0.7196, and 0.2734 with absolute values of 0.9698, 0.7196, and 0.2734, respectively. This implies that the linearized system is stable. If we set initial conditions for all variables to 0.1 and run the simulation for thousand samples, the simulation indicates that the nonlinear system is also stable with respect to the origin. This is depicted in Fig. 4.9 where the second variable of h is plotted versus the first variable, and in Fig. 4.10 where the fourth variable of h is plotted versus the third variable. In both figures, only last 30 samples are depicted.

Interesting enough if we change the initial conditions for all variables to be a bit further away from the origin, that is, to 0.2 we get a completely different behavior. In Fig. 4.11, the second variable of h is plotted versus the first variable, and in Fig. 4.12 the fourth variable of h is plotted versus the third variable. In both figures, only last 50 samples are plotted. The simulation is run for a longer time, that is, 2×10^5 samples.

These two figures show that the nonlinear system is not convergent with respect to the origin when starting from initial conditions at 0.2. What is also intriguing and interesting, is that the nonlinear system does not seem to behave as chaotic but rather experiences something that resembles a periodic behavior instead. This is very interesting and also very illustrative since in the case of a system model with dimension eight, when the linear model is stable, the nonlinear model and the corresponding linear model exhibit different dynamic behaviors from the stability point of view.

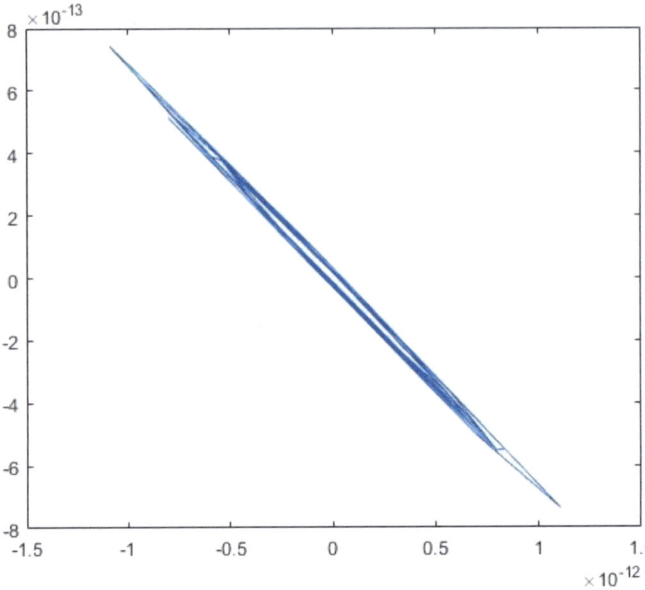

Fig. 4.10 Last 30 samples of h_4 versus h_3 for the eight dimensional example run for thousand samples starting from the initial conditions equal to 0.1

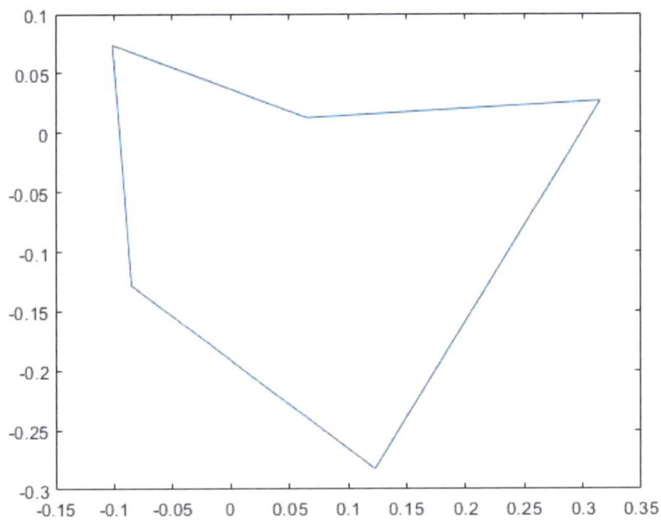

Fig. 4.11 Last 50 samples of h_2 versus h_1 for the eight dimensional example run for 2×10^5 samples starting from the initial conditions equal to 0.2

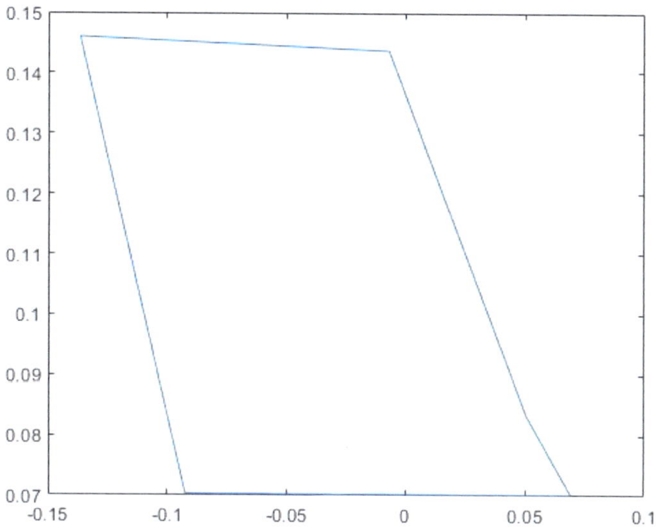

Fig. 4.12 Last 50 samples of h_4 versus h_3 for the eight dimensional example run for 2×10^5 samples starting from the initial conditions equal to 0.2

To conclude this section, we have shown by a set of representative simulations that there are indications that the "edge of chaos" for the nonlinear autonomous long short-term memory neural network model matches very accurately the boundary of stability of its linearized model, when an equilibrium is at zero. But we have also shown in the case when the system degree is increased to eight, that this matching is not perfect.

References

1. N. E. Barabanov and D. V. Prokhorov. Stability analysis of discrete-time recurrent neural networks. *IEEE Transactions on Neural Networks*, vol. 13, no. 2, pp. 292–303, 2002.
2. N. Bertschinger and T. Natschläger. Real-time computation at the edge of chaos in recurrent neural networks. *Neural Computation*, vol. 16, no. 7, pp. 1413–1436, 2004.
3. S. A. Deka, D. M. Stipanović, B. Murmann, and C. J. Tomlin. Global Asymptotic Stability and Stabilization of Long Short-Term Memory Neural Networks with Constant Weights and Biases. *Journal of Optimization Theory and Applications*, vol. 181, pp. 231-243, 2019.
4. S. A. Deka, D. M. Stipanović, B. Murmann, and C. J. Tomlin. Long Short-Term Memory Neural Network Stability and Stabilization using Linear Matrix Inequalities. *Proceedings of the 2019 IEEE International Symposium on Circuits and Systems*, Sapporo, Japan, 2019.
5. A. Graves. Generating sequences with recurrent neural networks. *arXiv*:1308.0850, 2013.
6. K. Greff, R. K. Srivastava, J. Koutník, B. R. Steunebrink, and J. Schmidhuber. LSTM: A search space odyssey. *IEEE Transactions on Neural Networks and Learning Systems*, vol. 28, no. 10, pp. 2222–2232, 2017.

7. J. Guckenheimer and P. Holmes. *Nonlinear Oscillations, Dynamical Systems, and Bifurcations of Vector Fields*. New York, NY, Springer-Verlag, 2002.
8. M. W. Hirsch, S. Smale, and R. L. Devaney. *Differential Equations, Dynamical Systems, and an Introduction to Chaos*, 3rd ed. Waltham, MA: Elsevier Inc., 2013.
9. S. Hochreiter and J. Schmidhuber. Long short-term memory. *Neural Computation*, 9(8):1735–1780, 1997.
10. D. W. C. Ho and G. Lu. Robust stabilization for a class of discrete-time non-linear systems via output feedback: the unified LMI approach. *International Journal of Control*, vol. 76, no. 2, pp. 105–115, 2003.
11. H. K. Khalil. *Nonlinear Systems*. 3rd Edition. Upper Saddle River, NJ: Prentice Hall, 2002.
12. J. P. LaSalle. *The Stability and Control of Discrete Processes*. New York, NY: Springer-Verlag, 1986.
13. T. Laurent and J. V. Brecht. A recurrent neural network without chaos. *Proceedings of the 2017 ICLR*. arXiv preprint arXiv:1612.06212, 2017.
14. D. M. Stipanović and D. D. Šiljak. Robust stability and stabilization of discrete-time non-linear systems: The LMI approach. *International Journal of Control*, vol. 74, pp. 873–879, 2001.
15. D. M. Stipanović, B. Murmann, M. Causo, A. Lekić, V. R. Royo, C. J. Tomlin, E. Beigne, S. Thuries, M. Zarudniev, and S. Lesecq. Some local stability properties of an autonomous long short-term memory neural network model. *Proceedings of the 2018 IEEE International Symposium on Circuits and Systems*, 2018.

5 Homotopy, Computation of Equilibria, and Long Short-Term Memory Neural Networks

As illustrated in the previous chapter for the case when there is an equilibrium at the origin, long short-term memory neural networks [8] have very complex and intriguing dynamic behaviors. Difference equations governing their dynamics are thus a great example of nonlinear discrete-time systems with a wide variety of different behaviors, ranging from stable to chaotic to unstable. One may stipulate that one of the reasons for the existence of such a variety of behaviors, at least partially, is due to the appearance of specific sigmoid functions and their concatenations with multiple weight matrices and biases, as they appear in the long short-term memory neural network dynamical model. Even most basic questions like the existence and the number of equilibria, are complex due to the transcendental nature of the corresponding fixed-point algebraic equation, which they are solutions to. Solving fixed-point nonlinear algebraic equations have been a long standing problem in mathematics [1, 5, 9]. One of the ways to approach the problem is to formulate and solve it using homotopy formulations [13, 14]. Essentially, this is done by creating a path for a, in general, multi-variable solution to traverse from a simple algebraic equation, or the one for which the solution is known, to the one which is a complex algebraic equation that is of interest to be solved.

In the previous chapter, we have studied some properties of the nonlinear autonomous long short-term memory neural network model with the bias gate set at zero, thus guaranteeing the existence of an equilibrium at zero. In this chapter we will remove this restriction and consider the full autonomous long short-term memory model. We will show that the boundary of stability of the linearized model even in this case is surprisingly accurate characterization of the "edge of chaos." As a reminder, chaos or chaotic behaviors are behaviors that cannot be classified as any other standard behaviors of dynamical systems [6, 7]. As mentioned more than a few times by now, the importance of the "edge of chaos" has been recognized earlier in connection to learning efficiency [3] of the recurrent neural networks. This is one of the reasons why we are interested in a specific dynamic behavior od a specific autonomous

D. Stipanović, *Difference Equations and Machine Learning*, Synthesis Lectures on Mathematics & Statistics, https://doi.org/10.1007/978-3-032-00910-4_5

nonlinear multi-variable difference equation, which again has important implications in machine learning and artificial intelligence.

The full nonlinear long short-term memory neural network discrete-time dynamic model, with the gate bias vector included, is governed by the following set of vector difference equations:

$$\begin{aligned} c(k+1) &= \varphi(\tilde{W}_f u(k) + W_f h(k) + \tilde{b}_f) \odot c(k) \\ &\quad + \varphi(\tilde{W}_i u(k) + W_i h(k) + \tilde{b}_i) \odot \tanh(\tilde{W}_g u(k) + W_g h(k) + \tilde{b}_g), \\ h(k+1) &= \varphi(\tilde{W}_o u(k) + W_o h(k) + \tilde{b}_o) \odot \tanh(c(k+1)), \end{aligned} \tag{5.1}$$

where the input or the reference vector is denoted as $u(\cdot)$ and k represents, as usual, discrete time instances taking, most commonly, nonnegative integer values. The weight matrices are denoted as $\tilde{W}_j$ and W_j and the bias vectors as $\tilde{b}_j$, where $j = \{f, i, o, g\}$. Tilde weight matrices are the ones multiplying the input signal. The reason for denoting bias vectors with tilde notation is that if the input signal, that is, either the training or to be classified signal is constant, that is,

$$u(k) \equiv u_c, \ \forall k,$$

then we can combine the constant input and tilde biases into autonomous-like biases, simply as

$$b_j = \tilde{W}_j u + \tilde{b}_j, \ \forall j \in \{f, i, o, g\}. \tag{5.2}$$

As an example of constant input signals is when either training or classification is done with image or symbol data sets. The result of this concatenation is the following nonlinear autonomous long short-term memory neural network model:

$$\begin{aligned} c(k+1) &= \varphi(W_f h(k) + b_f) \odot c(k) \\ &\quad + \varphi(W_i h(k) + b_i) \odot \tanh(W_g h(k) + b_g), \\ h(k+1) &= \varphi(W_o h(k) + b_o) \odot \tanh(c(k+1)). \end{aligned} \tag{5.3}$$

Discrete-time model in Eq. (5.3) is a result of combining Eqs. (5.1) and (5.2). Again, studying this nonlinear model may have dual purpose, depending on what is of our interest. One is to study properties of autonomous long-short term neural networks and the other is to analyze their use with constant inputs, either for the purpose of training or classification. Let us note here that the application to constant data sets is not of great importance perse given that the long short-term neural networks are mainly used for time-varying datasets. But then the motivation for the validation of analyzing this model stems from multiple reasons, where we will mention two of them. One is, again, the recurrent networks perform well when their corresponding autonomous models operate at the "edge of chaos" and the other stems from one particular property of discrete-time control problem formulations. This property is that discrete-time control problem formulations, if considered on finite time-horizons that progress in time like in the model predictive control formulations, can be transformed into a set of time-invariant systems within the corresponding time horizon. In this way, dynamical discrete-time control problem formulations are converted into a number of algebraic problem formulations that are to be solved in a sequence. By doing so, the time dependence is removed at least in terms of the problem formulation.

In the full autonomous discrete-time formulation, the discrete-time dynamic model in Eq. (5.3) does not have, in general, an equilibrium at the origin when the gate bias vector b_g is not equal to zero. The computation of nonzero, that is, non-trivial equilibria is not an easy task since the equilibria in this case are the solutions to the corresponding transcendental algebraic equations. One important aspect of the computation of equilibria is that the numerical value for each equilibrium is a necessary ingredient in studying properties of the corresponding nonlinear system in the vicinity of the equilibrium. This is because the analysis is based on the linearized models where the Jacobians are to be evaluated at the corresponding equilibria. One way, which is also known as a standard and well-known formulation for computing the equilibria, is to use a path-following or homotopy approach. Again, this approach is based on creating a path between a trivial to solve algebraic equation or the one with a known solution, and a complex and difficult to solve equation, for which we would like to compute a solution. This is also related to a well know technique of embedding, where one particular problem is embedded into a family of related problems. More details on the homotopy formulations are provided in the following section.

5.1 Homotopy Approach for Computing Equilibria of Long Short-Term Memory Models

As shown earlier and in [12], long short-term memory neural networks without the gate bias vector b_g have an equilibrium at the origin. When bias vector b_g is nonzero, then the computation of the corresponding nonzero equilibria is related to solving transcendental fixed-point algebraic equations. Since we know an equilibrium when $b_g = 0$ is set to zero to be at the origin, one idea is to create a homotopy with a path that originates at the origin as a solution to the fixed-point algebraic equation when the gate bias vector is zero, to a solution of the fixed point algebraic equation with nonzero gate bias vector [2]. The motivation to pick the starting algebraic equation to be the one corresponding to the autonomous long-short term memory neural network model with no gate bias vector, rather then a simple one like a trivial linear equation, is that the path itself may be easier to follow given that the initial and final algebraic equations are very similar and of the same structure.

5.1.1 Homotopy Formulations

The homotopy or path-following approach is a simple concept of creating a path between an algebraic equation that has a simple or a known solution and the one that needs to be solved, which is furthermore difficult to solve. One representative example, which is of our interest, would be an equation belonging to the class of nonlinear transcendental algebraic equations. For more details on homotopy and path-following formulations and related numerical methods to solve these algebraic equations, we refer to [14]. Again for our specific purpose, we will use this approach to solve for nontrivial, that is, nonzero equilibria of the long short-term memory neural network model (5.3) where the gate bias vector is not equal to zero.

In technical terms, the basic idea of the homotopy approach considers two functions

$$F, G : \mathbb{R}^n \to \mathbb{R}^m,$$

which we will assume to be continuous, unless stated otherwise. Here we assume that $F(x) = 0$ is an algebraic equation that either has a simple or a known solution and $G(x) = 0$ is the one we would like to solve. A homotopy concatenation involving these two functions is simply another continuous function

$$H : [0, 1] \times \mathbb{R}^n \to \mathbb{R}^m$$

where we set

$$\begin{aligned} H(0, x) &= F(x), \\ H(1, x) &= G(x). \end{aligned}$$

To accomplish the computation of nonzero equilibria objective, let us first consider a simple linear homotopy which represents a convex combination of the two functions $F(x)$ and $G(x)$, that is,

$$H(t, x) = (1 - t)F(x) + tG(x) \tag{5.4}$$

where t is a parameter or an independent variable, dependent on a treatment, such that $t \in [0, 1]$.

Again, our goal is to solve the following algebraic equation:

$$F(x) = 0 \tag{5.5}$$

assuming that $G(x) = 0$ has a known solution $x_0 \in \mathbb{R}^n$, that is,

$$G(x_0) = 0.$$

Then, the homotopy or path-following formulation is to solve a whole family of the following algebraic equations:

$$H(t,x) = (1-t)G(x) + tF(x) = 0, \forall t \in [0,1], \tag{5.6}$$

where again $H(0,x) = G(x)$ and $H(1,x) = F(x)$ for all $x \in \mathbb{R}^n$. This is a very well known approach in mathematics where one particular problem we are interested in solving, that is, $F(x) = 0$, is embedded in a whole set or family of solutions of the homotopy equation for all $t \in [0,1]$. This is essentially a creation of a family of problems with respect to a parameter (or later treated as a variable) t, which varies continuously from zero to one. This is a known technique in mathematics where analyzing a whole family of problems provides more information about one particular problem, which we are interested in solving.

Now, following standard techniques, let us define the set of solutions to the homotopy equation as follows (by adopting the notation from [14]):

$$H^{-1} = \left\{(t,x) \in [0,1] \times \mathbb{R}^n : H(t,x) = 0\right\}. \tag{5.7}$$

Then, one of the questions that is related to why this formulation is proposed in the first place, is if we can find a relationship between t as an independent variable and x as a dependent variable such that the function $x(t)$ exists, and it is, for example, at least continuous. This function $x(t)$ represents the family of solutions of the homotopy equation as t varies from zero to one. By utilizing the implicit function theorem as Path Theorem 1.4.2 in [14], if mapping $H(\cdot,\cdot)$ is continuously differentiable, and for every $y = (t,x) \in H^{-1}$, the Jacobian $\frac{\partial H(y)}{\partial y}$ is of full rank, then H^{-1} consists only of continuously differentiable paths.

For the overall homotopy Jacobian

$$\frac{\partial H(y)}{\partial y},$$

when expressed in terms of variables t and x, considered for this formulation as independent variables, we use the following notation adopted again from [14]:

$$H'(t,x) = \left[H'_x(t,x), \frac{\partial H(t,x)}{\partial t} \right], \text{ where } H'_x(t,x) = \frac{\partial H(t,x)}{\partial x}. \tag{5.8}$$

As it has been indicated in [14], the analysis in terms of $x(t)$ can be limiting and a less conservative approach is to assume another independent variable τ to be considered as a distance of travel on the path. That is, if we denote

$$y(\tau) = (x(\tau), t(\tau)) \tag{5.9}$$

then the vector $y(\tau)$ would be on the path if

$$y(\tau) \in H^{-1} \tag{5.10}$$

Next, if $y(\tau)$ describes a path then

$$H(y(\tau)) = 0 \Rightarrow \frac{\partial H(y)}{\partial y}\dot{y} = 0. \tag{5.11}$$

The result is that we get the following set of differential equations, which is referred to as the *basic differential equations* [14]:

$$\dot{y}_i = (-1)^i \det\left(H'_{-i}(y)\right), \quad i = 1, \cdot, n, n+1 \tag{5.12}$$

where $\det(\cdot)$ represents a determinant of a square matrix as an argument, $H'_{-i}(y)$ is the Jacobian of H with the i-th column deleted. Now, if mapping H is continuously differentiable and regular, by the Basic Differential Equations Theorem (Theorem 2.1.1 in [14]), the solution of the basic differential equations starting from an initial condition $y_0 = y(p_0)$, which is a solution to the homotopy, is unique and defines a path in H^{-1}.

It is of interest to note that by setting $t = \tau$ we get the following set of differential equations:

$$\frac{\partial H(x(t), t)}{\partial x}\frac{dx(t)}{dt} = -\frac{\partial H(x(t), t)}{\partial t}, \quad x(0) = x_0. \tag{5.13}$$

where $x(\cdot)$ is an integral curve of (5.13) with $x(1)$ being a solution to our desired set of algebraic equations, that is,

$$H(1, x(1)) \equiv F(x(1)) = 0. \tag{5.14}$$

Another interesting feature of the basic differential equations is that they need not be unique. As a matter of fact, if a notion of so-called "velocity" is expressed in terms of a continuous scalar function $q(\cdot)$, we can reformulate the basic differential equations as follows:

$$\dot{y}_i = q(\tau)(-1)^i \det\left(H'_{-i}(y)\right), \quad i = 1, \cdot, n, n+1 \tag{5.15}$$

which solutions will also create a homotopy path [14], that is, $y(\tau) \in H^{-1}$. Trivially, if $q(\tau) \equiv 1$ then we get the set of basic differential equations, while $q(\tau) \equiv -1$ would result in a reverse movement on the path. In general, $q(\tau)$ can be used for numerical computation purposes to generate how fast the solutions are moving on the path and thus there is a motivation to refer to it as "velocity," following again, notations adopted in [14].

5.1.2 Homotopy Approach to Compute Nonzero Equilibria of Long Short-Term Memory Neural Networks

When applying homotopy formulation to computation of nontrivial, that is, nonzero equilibria of the long short-term neural network autonomous model, we will utilize the fact that there is an equilibrium at zero when the gate bias vector equals zero. This will be the algebraic equation from which the homotopy path will emanate from. Therefore, let us consider the nonlinear autonomous long short-term neural network model in Eq. (5.3) and let

$$h(k) = z(k) + z_c, \tag{5.16}$$

where $z(\cdot) \in \mathbb{R}^n$ is a new variable and $z_c \in \mathbb{R}^n$ such that

$$W_g z_c + b_g = 0. \tag{5.17}$$

In order for Eq. (5.17) to be feasible, the bias vector b_g needs to be in the range of columns of W_g, which means that it can be represented as a linear combination of the columns of W_g. Obviously this is always satisfied when W_g is nonsingular. This is a very mild requirement and only if W_g is singular one would need to check if b_g is in the range of columns of the gate weight matrix. If W_g is nonsingular then from Eq. (5.17), we can simply compute the following:

$$z_c = W_g^{-1} b_g.$$

But given that the choice of W_g is up to the designer, the assumption of its nonsingularity, obviously considering the quadratic case, or that b_g belongs to its column spanned space, in general case, is very mild and should easily be accommodated by the designer. At this point let us also remind the reader that the weight matrices and bias vectors are essentially control variables to be computed or determined by the designer, and are the result of the corresponding training. Therefore, assuming that Eq. (5.17) is feasible, we can use the change of variables in Eq. (5.16) to rewrite (5.3) as follows:

$$\begin{aligned} c(k+1) &= \varphi(W_f z(k) + \bar{b}_f) \odot c(k) \\ &\quad + \varphi(W_i z(k) + \bar{b}_i) \odot \tanh(W_g z(k)), \\ z(k+1) &= \varphi(W_o z(k) + \bar{b}_o) \odot \tanh(c_{k+1}) - z_c. \end{aligned} \tag{5.18}$$

The modified bias vectors are again defined as

$$\bar{b}_j = b_j + W_j z_c, \text{ for } j \in \{f, i, o\}.$$

Next, to simplify the notation, let us combine the state variables into one vector, that is, let us define the overall state vector as the following:

$$x(k) := [c(k)^T, z(k)^T]^T \tag{5.19}$$

and denote the corresponding constant "shift" vector as

$$x_c := [0^T, -z_c^T]^T. \tag{5.20}$$

Now, with the change of variables and by adopting the new notation, Eq. (5.18) can be rewritten in a simple form as follows:

$$x(k+1) = F(x(k)) + x_c, \tag{5.21}$$

where

$$
\begin{aligned}
F(x) &= F\left(\begin{bmatrix} c \\ z \end{bmatrix}\right) \\
&= \begin{pmatrix} F_1(c, z) \\ F_2(c, z) \end{pmatrix} \\
&= \begin{pmatrix} \varphi(W_f z + \bar{b}_f) \odot c + \varphi(W_i z + \bar{b}_i) \odot \tanh(W_g z) \\ \varphi(W_o z + \bar{b}_o) \odot \tanh(F_1(c, z)) \end{pmatrix}.
\end{aligned} \tag{5.22}
$$

At this point we can formulate the following linear homotopy that fits our problem of computing nontrivial equilibria of the nonlinear long short-term memory neural network autonomous model:

$$
\begin{aligned}
H(t, x) &= (1 - t)(x - F(x)) + t(x - F(x) - x_c) \\
&= (x - F(x)) - t x_c = 0, \quad t \in [0, 1].
\end{aligned} \tag{5.23}
$$

Associated with this formulation we have that $H(t, x)$ is continuously differentiable and

$$
H(0, 0) = -F(0) = 0 \tag{5.24}
$$

with $H \in C^2$, that is, $H(\cdot, \cdot)$ is continuously differentiable. We also compute the Jacobian and partial derivatives of $H(t, x)$ as follows:

$$
H' = \left[I - \frac{\partial F}{\partial x}, -x_c \right], \; H'_x = I - \frac{\partial F}{\partial x}, \; H'_t = -x_c. \tag{5.25}
$$

So in this specific case, Eq. (5.11) turns into the following expression:

$$
\left(I - \frac{\partial F}{\partial x} \right) . \frac{dx}{d\tau} - x_c \frac{dt}{d\tau} = 0, \quad (t(0), x(0)) = (0, 0) \tag{5.26}
$$

which is a vector differential equation with specified initial conditions. Furthermore, if

$$
\frac{\partial H}{\partial x} = I - \frac{\partial F}{\partial x}
$$

is nonsingular, that is, of full rank, Eq. (5.13) would have a unique trajectory $(t, x(t)) \in H^{-1}$ emanating from the origin, and taking the following form:

$$
\left(I - \frac{\partial F}{\partial x} \right) . \frac{dx}{dt} = x_c \;\; \forall t \in [0, 1], \; x(0) = 0. \tag{5.27}
$$

The solution to this differential equation will start from $x = 0$ at $t = 0$ and terminate at a nontrivial equilibrium of (5.3) at $t = 1$, which is what we are interested in computing.

If of interest, there is an option to address the singularity of

$$
I - \frac{\partial F}{\partial x},
$$

since the initial algebraic equation $x - F(x) = 0$ has a solution at zero. Thus, it can be modified to the more general form

$$ax + bF(x) = 0$$

for any two real numbers a and b as long as $ab \neq 0$. Then, the linear homotopy becomes

$$\begin{aligned} H(t, x) &= (1-t)(ax + bF(x)) + t(x - F(x) - x_c) \\ &= (ax + bF(x)) + t((1-a)x \\ &\quad -(1+b)F(x) - x_c) = 0, \ t \in [0, 1]. \end{aligned} \tag{5.28}$$

and correspondingly

$$\begin{aligned} H'_x &= (a + (1-a)t)I + (b - (1+b)t)\frac{\partial F}{\partial x}, \\ H'_t &= (1-a)x - (1+b)F(x) - x_c. \end{aligned} \tag{5.29}$$

By regrouping the terms differently, the partial derivatives with respect to t and x, can be computed as follows:

$$\begin{aligned} H'_x &= ((1-t)a + t \cdot 1)I + ((1-t)b + t \cdot (-1))\frac{\partial F}{\partial x}, \\ H'_t &= (1-a)x - (1+b)F(x) - x_c. \end{aligned} \tag{5.30}$$

From the expression for H'_x in Eq. (5.30), the coefficient multiplying the identity matrix varies from a to one, and the one multiplying $\frac{\partial F}{\partial x}$ varies from b to minus one. The modification of the initial algebraic equation with a solution at the origin still results in $H'_x(1, x) = I - \frac{\partial F}{\partial x}$, which in this case is only for the single and final value of t, that is, when $t = 1$, and not for the full path.

Homotopy path-following differential equations have to be solved, in general, using numerical methods [4, 11]. But given there is a matrix multiplying the derivative of the dependent variables, as evident from Eqs. (5.13) and (5.27), these differential equations do belong to a more general category of differential-algebraic equations and thus need to be treated with particular attention and correspondingly [10].

5.2 Nonzero Equilibria Model Simulations

In this section we illustrate dynamic behaviors of the nonlinear autonomous long short-term memory neural networks with equilibria that are away from zero. We use a homotopy path that starts from the fixed-point equilibria equation [1, 9] with the bias gate vector equal to zero and ends at the fixed-point equation with a nonzero gate bias vector. This way, a path between the origin and a nonzero equilibrium is created via the corresponding path-following family of nonlinear algebraic equations. The simulation examples range from clearly stable to boundary stable, proceeding to a boundary unstable case, and end with a clearly unstable simulation example. Given we are targeting exactly these four scenarios, finding numerical values for the entries of weight matrices and bias vectors had to be done through a "trial and error" set of attempts. This was especially the case for the scenarios when some of the equilibria of the linearized model are near the boundary of stability. What will be shown by these simulations is that the outcomes match the experience we had with the nonlinear models with an equilibrium at the origin, that is, the boundary of stability for the linearized systems in the case of nonzero equilibria appears to be again remarkably accurate to predict the behaviors of the nonlinear model. The boundary of stability also appears to match the boundary of chaos, at least in simulations.

5.2.1 A Clearly Stable Simulation Example

The first simulation example is the one where the linearized model is clearly stable, meaning the eigenvalues of the linearized model are comfortably inside the unit circle. Therefore, the maximal absolute value among the eigenvalues is chosen to be $\lambda_M = 0.88$. The problem is four dimensional, where again this means that there are two h variables and two c variables. The numerical values for the entries of the weight matrices are the following:

$$W_i = \begin{bmatrix} -1 & -4 \\ -3 & -2 \end{bmatrix}, \quad W_o = \begin{bmatrix} -1 & -6 \\ 6 & -9 \end{bmatrix},$$
$$W_f = \begin{bmatrix} -3 & 6 \\ 0 & -6 \end{bmatrix}, \quad W_g = \begin{bmatrix} -3.5 & -9.51 \\ 6.51 & 9 \end{bmatrix}.$$

The numerical values for the two entries of the nonzero gate bias vector are chosen as follows:

$$b_g = \begin{bmatrix} 1.25 \\ 1.55 \end{bmatrix}$$

and the other three bias vectors are set to zero. After running the homotopy differential equation starting from the fixed-point algebraic equation corresponding to the long short-term memory autonomous dynamic model with the same weight matrices but with the gate bias vector set to zero, we obtain the following nonzero equilibrium:

$$\begin{bmatrix} h_e \\ c_e \end{bmatrix} = \begin{bmatrix} 5.76 \times 10^{-4} \\ 0.1311 \\ 1.84 \times 10^{-3} \\ 0.6270 \end{bmatrix}$$

where the notation indicates that the state variables are organized as follows:

$$\begin{bmatrix} h \\ c \end{bmatrix} = \begin{bmatrix} h_1 \\ h_2 \\ c_1 \\ c_2 \end{bmatrix}. \tag{5.31}$$

This order of state variables will remain the same for all simulations in this section. The eigenvalues of the linearized model are computed as

$$\lambda = \begin{bmatrix} -0.3554 + 0.8066j \\ -0.3554 - 0.8066j \\ 0.5744 \\ 1.535 \times 10^{-16} \end{bmatrix}$$

where

$$\lambda = \begin{bmatrix} \lambda_1 \\ \lambda_2 \\ \lambda_3 \\ \lambda_4 \end{bmatrix},$$

with the maximal absolute value

$$\lambda_M = \max_{i \in \{1,2,3,4\}} \{\lambda_i\}. \tag{5.32}$$

In this case, the maximal absolute value among the eigenvalues has been computed using Eq. (5.32) to be $\lambda_M = 0.88$. This numerical value was mentioned earlier yet to determine all of the reported numerical values some "trial and error" iterations had to be done. This λ_M value is well within the unit circle and thus implies clear stability of the linearized model.

Given that the linearized system is clearly stable and that the equilibrium is not too far from the origin we test the nonlinear long short-term memory neural network's behavior with initial conditions that are close to the unit box. Therefore, we initiate the simulation at 0.9 for each of the state variables, that is,

$$\begin{bmatrix} h(0) \\ c(0) \end{bmatrix} = \begin{bmatrix} 0.9 \\ 0.9 \\ 0.9 \\ 0.9 \end{bmatrix}.$$

The simulation is run for 50 samples and only last 10 are depicted. These two particular numbers of samples have been chosen as sufficient and appropriate for the illustration. In Fig. 5.1, state variable h_2 is plotted versus state variable h_1, where the plot itself indicates fast convergence to the equilibrium. Again, for the illustration purpose, and because of the fast convergence, only last 10 samples are plotted.

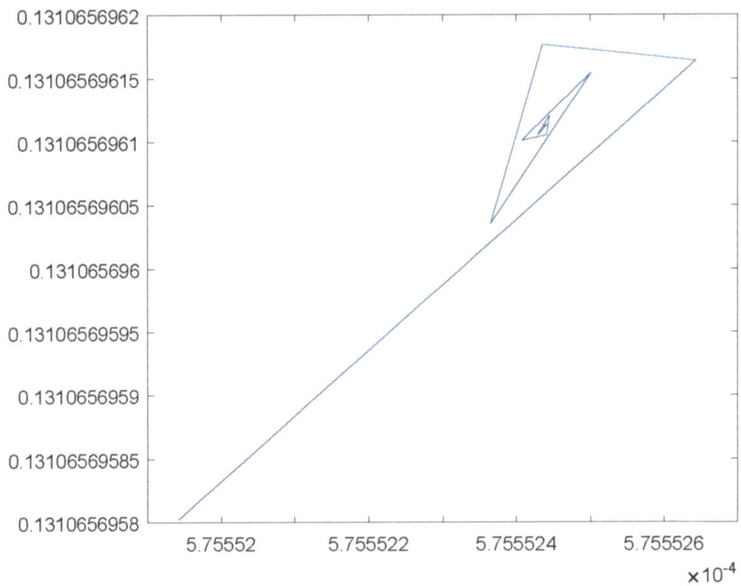

Fig. 5.1 Last ten samples of h_2 versus h_1 for the stable simulation example when $\lambda_M = 0.88$

The same simulation is also depicted in Fig. 5.2, where the only difference is that variable c_2 is plotted versus variable c_1. In conclusion, we can observe that for this choice of weight matrices and bias vectors, resulting in a nonzero equilibrium, the nonlinear long short-term memory neural network model produces a stable trajectory, which matches the stability property of the linearized model.

5.2.2 Close to Boundary of Stability yet a Stable Simulation Example

In this subsection, we consider an example where the maximal absolute value among the eigenvalues of the linearized system is close to one yet still all of the eigenvalues are inside the unit circle. The maximal absolute value among the eigenvalues in this example is chosen as $\lambda_M = 0.99$. The numerical values for the elements of the weight matrices are chosen as follows:

$$W_i = \begin{bmatrix} -1 & -4 \\ -3 & -2 \end{bmatrix}, \quad W_o = \begin{bmatrix} -1 & -6 \\ 6 & -9 \end{bmatrix}$$
$$W_f = \begin{bmatrix} -2 & 6 \\ 0 & -6 \end{bmatrix}, \quad W_g = \begin{bmatrix} -0.5 & -4.51 \\ 4.51 & -5 \end{bmatrix}$$

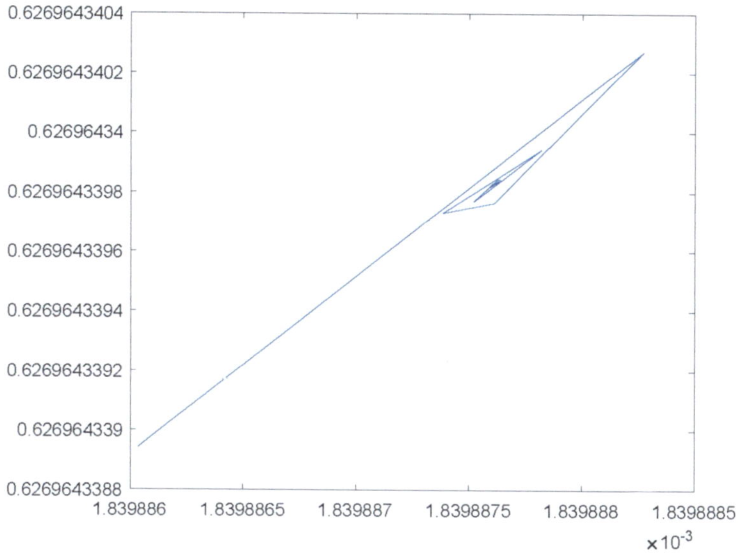

Fig. 5.2 Last ten samples of c_2 versus c_1 for the stable simulation example when $\lambda_M = 0.88$

The nonzero gate bias vector entries are chosen as follows:

$$b_g = \begin{bmatrix} 0.9 \\ -0.20 \end{bmatrix}.$$

while the other bias vectors are again set to zero. This choice of weight matrices and bias vectors results in a nonzero equilibrium located at

$$\begin{bmatrix} h_e \\ c_e \end{bmatrix} = \begin{bmatrix} 0.1488 \\ 0.0603 \\ 0.4198 \\ 0.1032 \end{bmatrix},$$

and the eigenvalues of the linearized model computed at this equilibrium turn out to be

$$\lambda = \begin{bmatrix} -0.6723 + 0.7289j \\ -0.6723 - 0.7289j \\ 3.82 \times 10^{-17} \\ 0.7545 \end{bmatrix}.$$

The maximal value among the eigenvalues is $\lambda_M = 0.99$ so that the linearized model is still stable with two eigenvalues placed close to the unit circle. Given that the expectation is that in this case the nonlinear long short-term neural network model would still be stable, we set the initial condition for each state variable to be 0.9, so away from the equilibrium and close to the boundary of the unit box. Therefore, we initiate the simulation at

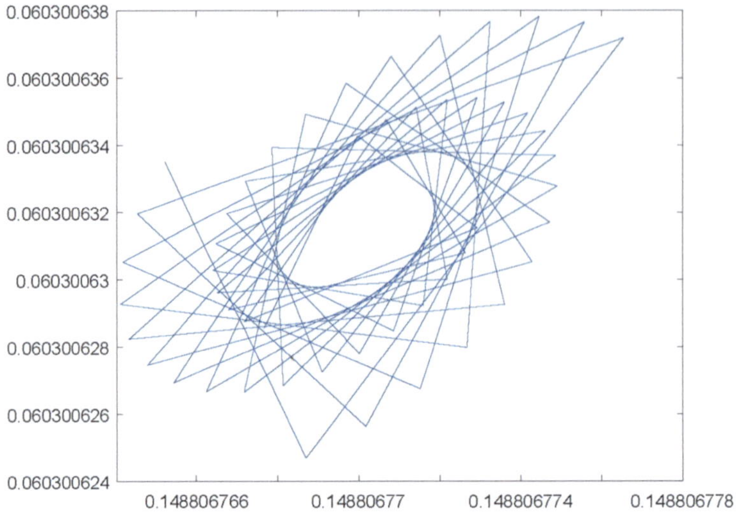

Fig. 5.3 Last 50 samples of h_2 versus h_1 for the close to boundary stable simulation example when $\lambda_M = 0.99$

$$\begin{bmatrix} h(0) \\ c(0) \end{bmatrix} = \begin{bmatrix} 0.9 \\ 0.9 \\ 0.9 \\ 0.9 \end{bmatrix}.$$

The simulation is run for one thousand samples and for the illustration purpose only last fifty samples in terms of state variable h_2 versus state variable h_1, are plotted in Fig. 5.3.

For the same simulation, state variable c_2 versus state variable c_1 is plotted in Fig. 5.4. The trajectory of the nonlinear system appears as stable and converges to the corresponding equilibrium, which again matches the stability property of the corresponding linear model.

5.2.3 Close to Boundary of Stability yet an Unstable Simulation Example

Now, we provide a simulation example when the eigenvalues "cross" the unit circle so that some of them are outside of the unit circle while they are placed in its vicinity. This implies that the linearized system is unstable.

The numerical values for the entries of the model weight matrices are chosen as

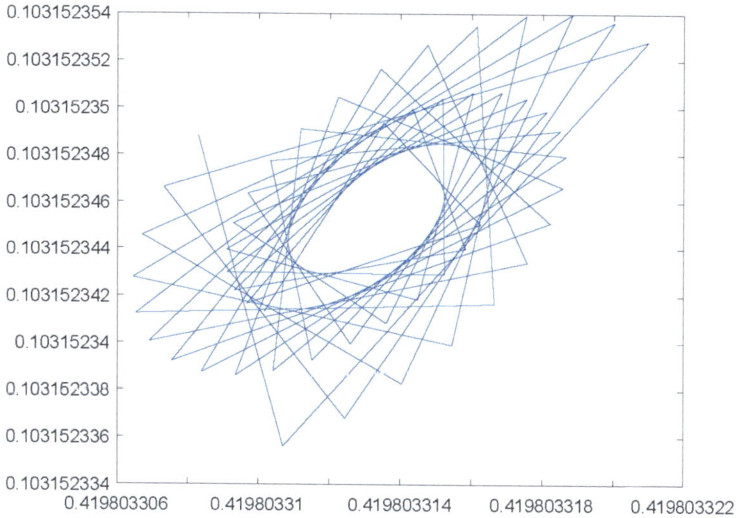

Fig. 5.4 Last 50 samples of c_2 versus c_1 for the close to boundary stable simulation example when $\lambda_M = 0.99$

$$W_i = \begin{bmatrix} -1 & -4 \\ -3 & -2 \end{bmatrix}, \quad W_o = \begin{bmatrix} -1 & -6 \\ 6 & -9 \end{bmatrix},$$

$$W_f = \begin{bmatrix} -2 & 6 \\ 0 & -6 \end{bmatrix}, \quad W_g = \begin{bmatrix} -0.5 & -4.51 \\ 4.51 & -5 \end{bmatrix}.$$

The numerical values for the elements of the gate bias vector are chosen as follows:

$$b_g = \begin{bmatrix} 1.0 \\ -0.26 \end{bmatrix}.$$

and all other bias vectors are set to zero.

This particular choice of weight matrices and bias vectors results in a nonzero equilibrium at

$$\begin{bmatrix} h_e \\ c_e \end{bmatrix} = \begin{bmatrix} 0.1617 \\ 0.0602 \\ 0.4656 \\ 0.0998 \end{bmatrix}.$$

Now, that the equilibrium is computed, we can compute the eigenvalues of the linearized model, which are obtained as follows:

$$\lambda = \begin{bmatrix} -0.6931 + 0.7334j \\ -0.6931 - 0.7334j \\ 8.19 \times 10^{-17} \\ 0.7759 \end{bmatrix}$$

with the maximal absolute value among them equal to $\lambda_M = 1.0091$. Given that the equilibrium has positive entries, one illustrative example is to start the simulation close to the unit box but on the negative side. Thus, the simulation is initiated at -0.9 for each state variable, that is,

$$\begin{bmatrix} h(0) \\ c(0) \end{bmatrix} = \begin{bmatrix} -0.9 \\ -0.9 \\ -0.9 \\ -0.9 \end{bmatrix}.$$

The simulation is run for ten thousand samples yet only last thirty are used for Fig. 5.5, where state variable h_2 is plotted versus state variable h_1. The plot indicates that the nonlinear long short-term memory dynamic model does not converge to an equilibrium, which is again matching the stability behavior of the linearized model.

Given that the trajectory does not converge, plotting c variables will not reveal anything new from the stability prospective but it does confirm that the equilibrium seems to be located within an ellipsoid-like shape created by the trajectory, which can be observed in Fig. 5.5. Therefore, the plot for c variables is provided in Fig. 5.6, and confirms the suggested enclosure of the computed equilibrium.

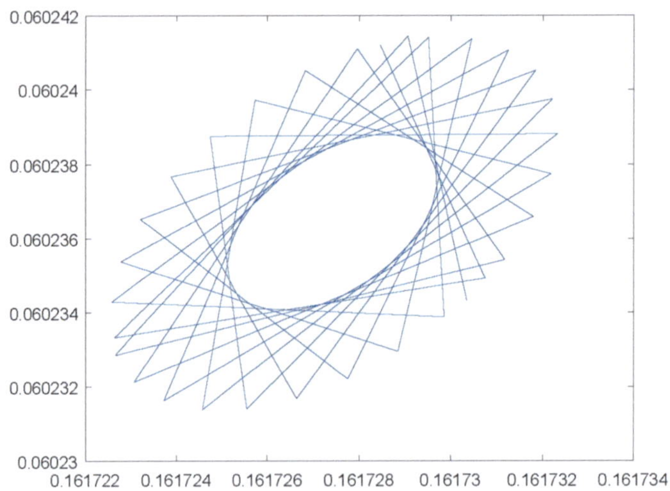

Fig. 5.5 Last 30 samples of state variable h_2 versus state variable h_1 when $\lambda_M = 1.0091$ and all state variables are initialized at -0.9

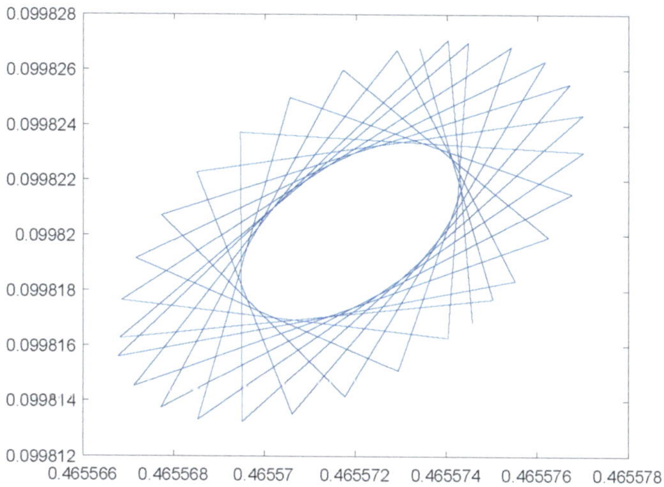

Fig. 5.6 Last 30 samples of state variable c_2 versus state variable c_1 when $\lambda_M = 1.0091$ and all state variables are initialized at -0.9

What is also interesting is to run another simulation when the state variable trajectories start from the values near the equilibrium. Thus, we choose the following initial condition:

$$\begin{bmatrix} h(0) \\ c(0) \end{bmatrix} = \begin{bmatrix} 0.1618 \\ 0.0603 \\ 0.4657 \\ 0.0999 \end{bmatrix}.$$

where we add 10^{-4} to each equilibrium entry, meaning we increase by one only the fourth decimal digit of each equilibrium entry. The simulation is again run for ten thousand samples, where only last thirty samples of the h variables are plotted in Fig. 5.7. Interesting enough, the trajectory again does not converge and the plots are practically the same, as when the vector initial condition was close to the unit box.

5.2.4 A Clearly Unstable Simulation Example

In this subsection, we provide the last homotopy related simulation example, when the linearized system is clearly unstable. In this particular case, the maximal absolute value among the eigenvalues is chosen to be $\lambda = 1.33$. When compared to the previous example, the eigenvalues with the maximal absolute value (that is, radius) are more distinctively away from the unit circle.

To push the eigenvalues away from the unit circle, the numerical values for the entries of the weight matrices have been changed to the following values:

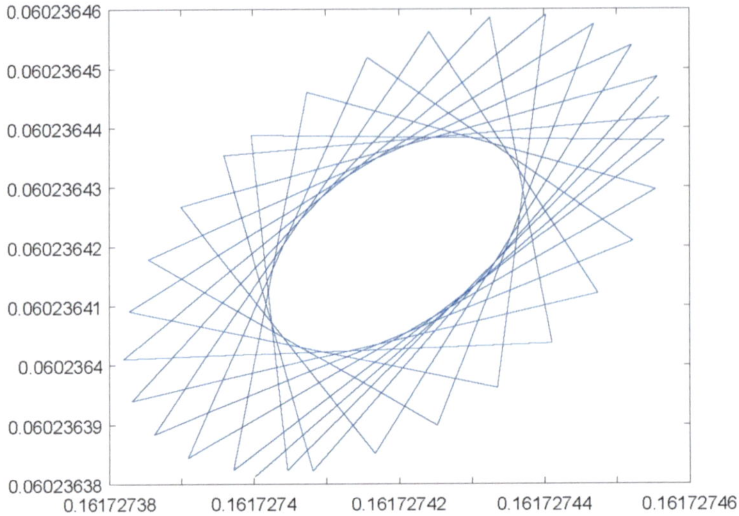

Fig. 5.7 Last 30 samples of state variable h_2 versus state variable h_1 when $\lambda_M = 1.0091$ and the initial condition is near the equilibrium

$$W_i = \begin{bmatrix} 1.18 & -3.04 \\ 7.08 & -1.08 \end{bmatrix}, \quad W_o = \begin{bmatrix} 8.78 & -4.09 \\ -3.97 & -3.34 \end{bmatrix},$$
$$W_f = \begin{bmatrix} -0.66 & -9.5 \\ 2.96 & 6.84 \end{bmatrix}, \quad W_g = \begin{bmatrix} -8.92 & 3.26 \\ -6.46 & -3.38 \end{bmatrix}.$$

The numerical values for the entries of the nonzero gate bias vector have been modified to the following:

$$b_g = \begin{bmatrix} -1.24 \\ -0.71 \end{bmatrix}.$$

where, like previously, the other bias vectors are set to zero.

The homotopy path computation approach provides the following nonzero equilibrium solution:

$$\begin{bmatrix} h_e \\ c_e \end{bmatrix} = \begin{bmatrix} -0.1017 \\ -0.0084 \\ -0.3557 \\ -0.0139 \end{bmatrix}$$

which can be used to compute the corresponding Jacobian of the linearized model. Then, the computation of the eigenvalues of the corresponding Jacobian provides the following numerical values:

$$\lambda = \begin{bmatrix} -0.9467 + 0.9401j \\ -0.9467 - 0.9401j \\ 0.6706 + 0j \\ -1.1817 \times 10^{-17} + 0j \end{bmatrix}$$

with the maximal absolute value among the eigenvalues is computed to be $\lambda_M = 1.33$, as indicated at the beginning of this subsection.

Given that the equilibria entries are all negative, the simulation is initialized close to the unit box yet on the positive side, that is,

$$\begin{bmatrix} h(0) \\ c(0) \end{bmatrix} = \begin{bmatrix} 0.9 \\ 0.9 \\ 0.9 \\ 0.9 \end{bmatrix}.$$

The simulation is run for ten thousand samples, and only last fifty samples are used to plot Fig. 5.8. The plotted trajectory clearly indicates that it does not converge toward the computed equilibrium, which is consistent with the linearized system being unstable.

The linearized system is clearly unstable in this case, and from Fig. 5.8 it can be observed that the closed curve formed by chaotic motions encloses the equilibrium, at least in terms of h variables. But like in the previously presented boundary unstable case, this is also true in terms of c variables, as depicted in Fig. 5.9. Interesting enough, in this case this enclosure is even more evident, arguably due to the system being more decisively unstable.

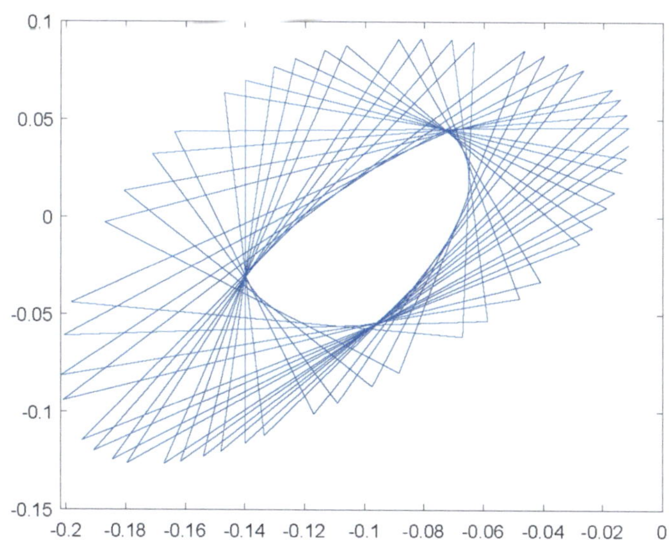

Fig. 5.8 Last 50 samples of state variable h_2 versus state variable h_1 when $\lambda_M = 1.35$ and all state variables are initialized at 0.9

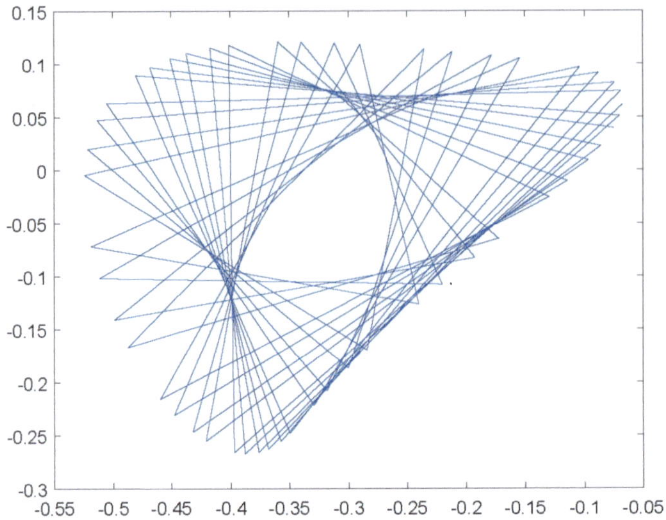

Fig. 5.9 Last 50 samples of state variable c_2 versus state variable c_1 when $\lambda_M = 1.33$ and all state variables are initialized at 0.9

Next, we run the simulation from another vector initial condition that is very close to the equilibrium. In light of this scenario, we choose the following vector initial condition:

$$\begin{bmatrix} h(0) \\ c(0) \end{bmatrix} = \begin{bmatrix} -0.1016 \\ -0.0083 \\ -0.3556 \\ -0.0138 \end{bmatrix}$$

where only the last decimal digit of each equilibrium entry is decreased by one, that is, 10^{-4} is added to each component, given that all entries are negative. The simulation is again run for ten thousand samples yet for the illustration purpose only last thirty were used for plotting Fig. 5.10.

The plots of the simulation starting near the edge of the unit box and the one starting near the equilibrium are, like in the previous case, practically identical.

In this section, the simulations were used to illustrate the use of homotopy path-following methodology to compute nonzero equilibria of the autonomous long short-term memory neural network model with nonzero values for the bias gate vector. This resulted in equilibria being placed away from the origin. Without any particular technical claims, it could be observed, like in the case when an equilibrium was placed at the origin, that in the case of nonzero equilibria, the nonlinear and linear models exhibit the same convergence and stability properties. This again, has been observed for the long short-term memory neural network models of dimension four.

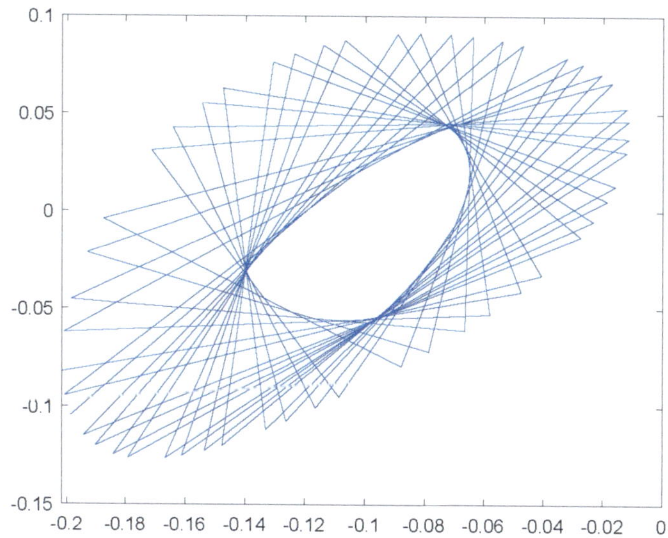

Fig. 5.10 Last 30 samples of state variable h_2 versus state variable h_1 when $\lambda_M = 1.33$ and the initial condition is near the equilibrium

References

1. R. P. Agarwal, M. Meehan, and D. O'Regan. *Fixed Point Theory and Applications*. Cambridge University Press, Cambridge, UK, 2004.
2. M. Amrouche, S. A. Deka, A. Lekić, V. R. Royo, D. M. Stipanović, B. Murmann, and C. J. Tomlin. Long Short-Term Memory Neural Network Equilibria Computation and Analysis. *Proceedings of the Workshop on Modeling and decision-making in the spatiotemporal domain, 32nd Conference on Neural Information Processing Systems (NIPS).* Montreal, Canada, 2018.
3. N. Bertschinger and T. Natschläger. Real-time computation at the edge of chaos in recurrent neural networks. *Neural Computation*, vol. 16, no. 7, pp. 1413–1436, 2004.
4. J. C. Butcher. *Numerical Methods for Ordinary Differential Equations*. John Wiley & Sons, Chichester, England, 2003.
5. B. P. Demidovich and I. A. Maron. *Computational Mathematics*. Mir Publishers, Moscow, 1981.
6. J. Guckenheimer and P. Holmes. *Nonlinear Oscillations, Dynamical Systems, and Bifurcations of Vector Fields*. New York, NY, Springer-Verlag, 2002.
7. M. W. Hirsch, S. Smale, and R. L. Devaney. *Differential Equations, Dynamical Systems, and an Introduction to Chaos*, 3rd ed. Waltham, MA: Elsevier Inc., 2013.
8. S. Hochreiter and J. Schmidhuber. Long short-term memory. *Neural computation*, vol. 9, no. 8, pp. 1735–1780, 1997.
9. L. V. Kantorovich and G. P. Akilov. *Functional Analysis*. Second Translated Edition, Pergamon Press, 1982.
10. P. Kunkel and V. Mehrmann. *Differential-algebraic Equations: Analysis and Numerical Solution*. European Mathematical Society, 2006.

11. J. D. Lambert. *Numerical Methods for Ordinary Differential Systems: The Initial Value Problem.* John Wiley & Sons, Inc., 1991.
12. D. M. Stipanović, B. Murmann, M. Causo, A. Lekić, V. Rubies Royo, C. J. Tomlin, E. Beigne, Thuries, Z. S., M., and S. Lesecq. Some local stability properties of an autonomous long short-term memory neural network model. *Proceedings of the 2018 IEEE International Symposium on Circuits and Systems (ISCAS)*, 2018.
13. L. T. Watson. A globally convergent algorithm for computing fixed points of C2 maps. *Applied mathematics and computation*, vol. 5, no. 4, pp. 297–311, 1979.
14. W. Zangwill and C. Garcia. *Pathways to Solutions, Fixed Points, and Equilibria.* Englewood Cliffs, NJ: Prentice-Hall, 1981.

6 Stability of Convex Combinations and Gated Recurrent Unit Neural Networks

Regarding recurrent neural networks, that is, dynamic neural networks, we have so far discussed properties of the long-short-term neural networks [8, 9, 11]. These properties have been shown as interesting and intriguing, while we could not do much about providing their technical characterizations. Therefore, in this chapter we will concentrate on a simplification of the long short-term memory neural networks that are known as gated recurrent neural networks [3]. These networks have also been very popular and proved as effective in various machine learning applications. However, the primary reasons for our interest in gated recurrent unit recurrent neural networks are twofold. On one hand, because of a particular simplification they can be linked to convex combinations of two discrete-time dynamical systems which enables us to provide more technical characterizations of their dynamic behaviors. On the other hand, since the gated recurrent unit neural network model is derived from the long short-term memory neural network model, it also exhibits interesting and complex dynamic behaviors as a nonlinear discrete-time dynamical system. The presentation and results in this chapter, recall and build upon the developments and results presented in [23].

6.1 Convex Combination Like Discrete-Time Autonomous Dynamical Systems

Dynamic properties of the gated recurrent unit neural networks [3] are represented by a set of nonlinear difference equations that are time-invariant, like in the case of the long short-term memory neural networks. Specifically, these difference equations have the following form:

D. Stipanović, *Difference Equations and Machine Learning*, Synthesis Lectures on Mathematics & Statistics, https://doi.org/10.1007/978-3-032-00910-4_6

$$
\begin{aligned}
h(k+1) = {} & \varphi_f(h(k)) \odot h(k) + (\mathbf{1}_n - \varphi_f(h(k))) \\
& \odot \tanh(W_g(\varphi_o(h(k)) \odot h(k)) + b_g), \qquad (6.1)\\
\text{where } \varphi_o(h(k)) = {} & \varphi(W_o h(k) + b_o) \text{ and } \varphi_f(h(k)) = \varphi(W_f h(k) + b_f)),
\end{aligned}
$$

where the presented form represents one layer of the network. Multiple layers would simply be represented as a concatenation of single layers, as if any other recurrent neural network has been used. The input signals are omitted for now, resulting in Eq. (6.1) being classified as autonomous and time-invariant nonlinear difference equations. Discrete time instances, which are assumed to be nonnegative integers, are as usual represented by k and $h(\cdot) = [h_1(\cdot), \ldots, h_n(\cdot)]^T$ is at the same time the n-dimensional state and output vector. Notice that in the case of gated recurrent unit neural networks there is no need to have c variables. Standard logistic function

$$\varphi(\cdot) = \frac{e^{(\cdot)}}{1 + e^{(\cdot)}}$$

is denoted in the same way and represents a sigmoid function that varies between zero and one. So it is a positive sigmoid function. The other sigmoid function used in the gated recurrent unit neural network dynamical model, is the hyperbolic tangent function, also defined in the standard way as

$$\tanh(\cdot) = \frac{e^{(\cdot)} - e^{-(\cdot)}}{e^{(\cdot)} + e^{-(\cdot)}}$$

where its image values range between minus one and one. Both logistic and hyperbolic tangent functions are applied element-wise, and $\odot$ represents the standard Hadamard product, as before. The n-dimensional column vector with all elements equal to one is again denoted as $\mathbf{1}_n$, that is, it represents the following constant vector:

$$\mathbf{1}_n = [1, \ldots, 1]^T \in \mathbb{R}^n. \qquad (6.2)$$

Weight matrices are denoted as W_j and bias vectors as b_j, where $j \in \{o, f, g\}$, with their entries assumed to be constant and thus the gated recurrent unit difference equations are time-invariant. Again, the weight matrices and bias vectors' numerical values are the result of an appropriate training so technically speaking are control parameters, yet given that at the end of the training they take constant values, the resulting difference equations are treated as time-invariant.

In order to emphasize that the gated recurrent unit difference equation dynamical model can be represented in a convex combination fashion, let us introduce the following function:

$$g(h) = \tanh(W_g(\varphi_o(h) \odot h) + b_g), \qquad (6.3)$$

and accordingly rewrite the overall gated recurrent unit difference equation (6.1) in the corresponding form as follows:

$$h(k+1) = \varphi_f(h(k)) \odot h(k) + (\mathbf{1}_n - \varphi_f(h(k))) \odot g(h(k)). \qquad (6.4)$$

There are two simple consequences, we can exploit, of the form presented in Eq. (6.4). The first one, is a simple rewriting as a set of coupled scalar difference equations in the following form:

$$h_i(k+1) = \varphi_{fi}(h(k))h_i(k) + (1 - \varphi_{fi}(h(k)))g_i(h(k)),\ i \in \{1, \ldots, n\}. \tag{6.5}$$

In this expression, we introduced the following function:

$$\varphi_{fi}(h) = \varphi(W_{fi}h + b_{fi}) \tag{6.6}$$

where W_{fi} denotes the i-th row of W_f and b_{fi} denotes the i-th element of b_f. This multiple scalar form explicitly illustrates that each scalar equation represents a convex combination. Thus, the overall set of difference equations can be understood as a connected set of scalar convex combinations. This is one of the intrinsic properties of the gated recurrent unit network dynamical model.

The other consequence is that the two discrete-time dynamical systems

$$h(k+1) = g(h(k))$$

and

$$h(k+1) = \varphi_f(h(k)) \odot h(k) + (\mathbf{1}_n - \varphi_f(h(k))) \odot g(h(k))$$

have the same equilibria for any function $g(\cdot)$. These equilibria are defined by the function $g(\cdot)$ itself, as solutions to the following vector algebraic equation

$$h^* = g(h^*). \tag{6.7}$$

Therefore, the equilibria, denoted as $h^* \in \mathbb{R}^n$ are fixed points of the $g(\cdot)$ map. This simple fact can be technically illustrated by the following:

$$\begin{aligned} h^* = \varphi_f(h^*) \odot h^* + (\mathbf{1}_n - \varphi_f(h^*)) \odot g(h^*) \Leftrightarrow\ & (\mathbf{1}_n - \varphi_f(h^*)) \odot h^* \\ & = (\mathbf{1}_n - \varphi_f(h^*)) \odot g(h^*) \\ \Leftrightarrow\ & h^* = g(h^*) \end{aligned} \tag{6.8}$$

where it has been used that the values of the logistic function are strictly larger than zero and strictly smaller than one, and therefore the values of the function and the values of the function subtracted from one, are nonzero.

6.2 Stability Analyses of a Class of Convex Combination Discrete-Time Dynamical Systems

The convex combination between two points essentially represents a set of all points on the line connecting those two points. But its influence goes far beyond its basic definition and has been used as a powerful concept in many mathematical definitions and formulations. One quite interesting and appealing is an application in modeling behaviors on a discontinuity surfaces for differential equations with discontinuous righthand sides [7, 17, 21]. The other application is related to decentralized consensus estimation algorithms [18]. As shown, this formulation has also been used in particular model dynamics like the gated recurrent unit model. It is important to note that the formulation itself offers some benefits when it appears on the righthand side of not only differential but difference equations too. To illustrate this line of thought, let us consider analyzing stability of a convex combination nonlinear discrete-time dynamical model, provided in Eq. (6.4). First, let us consider the scalar case, which is simply

$$h(k+1) = \varphi_f(h(k))h(k) + (1 - \varphi_f(h(k)))g(h(k)), \tag{6.9}$$

and is clearly in the form of a time-varying convex combination between a simple linear system $h(k+1) = h(k)$ and a more complicated nonlinear system $h(k+1) = g(h(k))$. Here, we use time-varying to denote that the convex combination parameter in this case is replaced by function $\varphi_f(h(k))$, while the gated recurrent unit neural network difference equation is being treated as time-invariant. For illustration purposes, assume that system $h(k+1) = g(h(k))$ is stable in the sense of Lyapunov with the corresponding convex Lyapunov function $v(\cdot)$ [13]. This assumptions means that

$$\Delta v(h(k)) = v(h(k+1)) - v(h(k)) = v(g(h(k))) - v(h(k)) \leq 0, \ \forall k. \tag{6.10}$$

Now let us apply the same Lyapunov function to the convex combination model in Eq. (6.9) and obtain

$$\begin{aligned} \Delta v(h(k)) &= v(h(k+1)) - v(h(k)) \\ &= v(\varphi_f(k)h(k) + (1 - \varphi_f(k))g(h(k))) - v(h(k)) \\ &\leq \varphi_f(k)v(h(k)) + (1 - \varphi_f(k))v((g(h(k))) - v(h(k)) \leq 0. \end{aligned} \tag{6.11}$$

Here we note again that in our opinion, the convexity is not a conservative assumption to impose on the Lyapunov function. For example, all standard p-norm based Lyapunov functions are convex.

We recall that the two discrete-time dynamical systems $h(k+1) = g(h(k))$ and $h(k+1) = \varphi_f(h(k)) \odot h(k) + (1 - \varphi_f(h(k))) \odot g(h(k))$ have the same equilibrium points given as solutions to (6.7). Therefore, to study stability with respect to equilibria of the convex combination system can be done with convex Lyapunov functions appropriate for the system $h(k+1) = g(h(k))$. Here we are using plural since a Lyapunov function has to be positive definite with respect to the corresponding equilibrium, and if there is more than one equilibrium then there should be more than one Lyapunov function, in general. Now, we will concentrate on local stability given that both long short-term memory and gated recurrent unit networks have unit box as an attractive and invariant set [22]. So pretty much all of the interesting and challenging dynamics happens within the unit box, which nevertheless does not simplify the complexity of the behaviors, especially the chaotic ones [10, 14]. Essentially, the complexity of the behaviors is contained within the unit box.

As shown, the overall difference equation can be written as n scalar difference equations of the form given in Eq. (6.5), which are themselves convex combinations. Now, we will show that the vector difference equation can be represented as a larger dimensional convex combination in the following "vertex" convex combination form [23]:

$$h(k+1) = \sum_{j=1}^{2^n} \alpha_j(h(k))G_j(h(k)). \tag{6.12}$$

where $G_j(h(k))$, $j \in \{1, \ldots, 2^n\}$ represent the vertices. To get the expressions for the vertices and convex combination coefficients in Eq. (6.12), let us first define an n-dimensional vector Q_j with binary entries, where each entry equals the binary number representation of $2^n - j$ and denote this formulation as

$$Q_j = \text{bin}(2^n - j). \tag{6.13}$$

The l-th entry of this vector is a binary number denoted as $Q_j(l)$. After introducing this notation, the components of Eq. (6.12) are defined as follows [23]:

$$\begin{aligned} &\bar{\varphi}_{jl}(h) = \varphi_l(h)^{Q_j(l)}(1-\varphi_l(h))^{1-Q_j(l)},\ \alpha_j(h) = \prod_{l=1}^{n} \bar{\varphi}_{jl}(h) \geq 0,\ \sum_{j=1}^{2^n} \alpha_j(h) = 1 \\ &G_j(h) = \left[G_{j1}(h), ..., G_{jn}(h)\right]^T,\ \ G_{jl}(h) = \begin{cases} h_l, & if\ Q_j(l) = 1 \\ g_l(h), & if\ Q_j(l) = 0 \end{cases}. \end{aligned} \tag{6.14}$$

This derivation also follows from a standard transformation of a box structure in Eq. (6.4) into a polytope or a vertex structure in Eq. (6.12) (see, for example, Sect. 8.3 in [1, 20]).

Now, we have all the ingredients to formulate the following theorem [23].

Theorem 6.1 *If there is a common convex Lyapunov function $v(h)$ that establishes stability of each system $h(k+1) = G_j(h(k))$ with respect to an equilibrium $g(h^*) = h^*$, then the overall system (6.4) is stable with respect to the same equilibrium.*

Proof The first thing to note is that if h^* is a fixed point of $g(h)$ then it is a fixed point of each $G_j(h)$ and thus an equilibrium of (6.12), by construction. The stability is implied by the convexity of $v(\cdot)$ and by using Eq. (6.12), that is,

$$
\begin{aligned}
v(k+1) \doteq v(h(k+1)) = \ & v\left(\sum_{j=1}^{2^n} \alpha_j(h(k))G_j(h(k))\right) \\
\leq & \sum_{j=1}^{2^n} \alpha_j(h(k))v\left(G_j(h(k))\right) \\
\leq & \sum_{j=1}^{2^n} \alpha_j(h(k))v(h(k)) = v(h(k)) \doteq v(k)
\end{aligned}
\tag{6.15}
$$

which completes the proof. □

We note that Theorem 6.1, in general, is applicable to scenarios where

$$
\mid g_i(h) \mid < 1, \ \ \forall i \in \{1, \dots, n\}, \tag{6.16}
$$

which is, for example, satisfied by both logistic and hyperbolic tangent sigmoid functions. For the scalar case, if this assumption is satisfied, then

$$
h_i(k+1) = \varphi_{fi}(h(k))h_i(k) + (1 - \varphi_{fi}(h(k)))g_i(h(k)) \tag{6.17}
$$

satisfies, the following property

$$
\mid h_i(k) \mid > 1 \Rightarrow \mid h_i(k+1) \mid < \mid h_i(k) \mid . \tag{6.18}
$$

and thus the solutions will be clearly bounded. This can be easily proved since

$$
|h_i(k)| > 1 > \mid g_i(h) \mid \tag{6.19}
$$

implies

$$
\begin{aligned}
\mid h_i(k+1) \mid \leq \ & \varphi_{fi}(h(k)) \mid h_i(k) \mid + (1 - \varphi_{fi}(h(k))) \mid g_i(h(k)) \mid \\
< \ & \varphi_{fi}(h(k)) \mid h_i(k) \mid + (1 - \varphi_{fi}(h(k))) \mid h_i(k) \mid \\
- \ & \mid h(k) \mid .
\end{aligned}
\tag{6.20}
$$

This means that every solution of each scalar difference equation will converge to the set $[-1, 1]$ and thus the vector solution will converge to the n-dimensional unit box $[-1, 1]^n$. The proof can be carried out for the overall system by using the following Lyapunov function

$$
v(h(k)) = \|h(k)\|_\infty = \max_i \{h_i(k)\}.
$$

In addition, we just need a reminder that by the governing fixed-point algebraic equations all equilibria are inside the unit box $[-1, 1]^n$.

What we have established so far, is the attractive and invariance properties of the unit box $[-1, 1]^n$ and that all equilibria must be inside this unit box. Again, the equilibria are solutions to, in general, transcendental equation $g(h^*) = h^*$. Assuming we can compute the equilibria, where an example is an equilibrium at the origin when the bias gate vector is zero in the long short-term memory model, we can use the linearization as the standard tool for analyzing local stability of the system around the corresponding equilibrium. The first step in doing so is the computation of the linearized model, that is, the Jacobian matrix. To streamline the presentation, let us use the following notation

$$\varphi_{fi}(h) = \varphi(a_i(h)), \text{ where } a_i(h) = W_{fi}h + b_{fi}. \tag{6.21}$$

Row vector W_{fi} is the i-th row of W_f and b_{fi} represents the i-th element of bias vector b_f. Furthermore,

$$F(h) = \varphi_f(h) \odot h + (1 - \varphi_f(h)) \odot g(h) \tag{6.22}$$

such that

$$F_i(h) = \varphi(a_i(h))h_i + (1 - \varphi(a_i(h)))g_i(h). \tag{6.23}$$

Now, we can start computing the Jacobian as

$$\begin{aligned} \frac{\partial F_i(h)}{\partial h} &= \varphi(a_i(h))(1 - \varphi(a_i(h)))h_i \frac{\partial a_i}{\partial h} + \varphi(a_i(h))e_i^T \\ &\quad -\varphi_i(a_i(h))(1 - \varphi(a_i(h)))g_i(h)\frac{\partial a_i}{\partial h} \\ &\quad +(1 - \varphi(a_i(h)))\frac{\partial g_i(h)}{\partial h} \\ &= \varphi(a_i(h))(1 - \varphi(a_i(h)))(h_i - g_i(h))W_{fi} \\ &\quad +\varphi(a_i(h))e_i^T + (1 - \varphi(a_i(h)))\frac{\partial g_i(h)}{\partial h}, \quad \forall i \in \{1, \dots, n\}, \end{aligned} \tag{6.24}$$

where e_i is a special n-dimensional column vector with all zeros except the i-th element being equal to one (thus the standard unit vector notation). The Jacobian needs to be evaluated at the corresponding equilibrium where

$$h_i^* = g_i(h^*), \ \forall i \in \{1, \dots, n\}.$$

Writing individual Eq. (6.24) in a compact form, results in the following:

$$\begin{aligned} \left.\frac{\partial F(h)}{\partial h}\right|_{h=h^*} &= D_1 + (I - D_1) \left.\frac{\partial g(h)}{\partial h}\right|_{h=h^*} \quad \text{where} \\ D_1 &= \text{diag}\{\varphi(a_1(h^*)), \dots, \varphi(a_n(h^*))\}. \end{aligned} \tag{6.25}$$

If we use a bit different notation where

$$D_2 = \text{diag}\{1 - \varphi(a_1(h^*)), \dots, 1 - \varphi(a_n(h^*))\}, \tag{6.26}$$

a different representation is obtained as follows:

$$\left.\frac{\partial F(h)}{\partial h}\right|_{h=h^*} = (I - D_2) + D_2 \left.\frac{\partial g(h)}{\partial h}\right|_{h=h^*} = I + D_2 \left(\left.\frac{\partial g(h)}{\partial h}\right|_{h=h^*} - I \right). \tag{6.27}$$

Equation (6.27) provides a relationship between the eigenvalues of

$$\frac{\partial F(h)}{\partial h}$$

and the eigenvalues of

$$\frac{\partial g(h)}{\partial h}.$$

After an equilibrium is computed and its value substituted in matrices in Eq. (6.27), rendering them constant values, they can be represented as follows:

$$J = \left.\frac{\partial F(h)}{\partial h}\right|_{h=h^*}, \tag{6.28}$$

$$L = \left.\frac{\partial g(h)}{\partial h}\right|_{h=h^*}, \tag{6.29}$$

$$D = D_2 = \mathrm{diag}\{1 - \varphi(a_1(h^*)), \ldots, 1 - \varphi(a_n(h^*))\}. \tag{6.30}$$

In conclusion, crucial issues in these derivations are the computation of and dependence on equilibria and the relationship between the Jacobian matrices J and L. There is not much that can be done with the equilibria computation when the equilibria are solutions to transcendental algebraic equations except using numerical computation, like path-following, or an alternative technical formulation option if possible, like the one formulated in the following lemma.

Lemma 6.1 *If D is a diagonal square matrix of dimension $n \times n$ such that all of its entries satisfy $0 < d_i < 1$, and L is an arbitrary square matrix with real entries of the same dimension such that $\|L\| < 1$, for any vector induced matrix norm, then $J = I + D(L - I)$ has all of its eigenvalues inside the unit circle, that is, it is Schur stable.*

Proof We start by writing the standard eigenvalue/eigenvector equation for an $n \times n$ matrix $D(L - I)$ as

$$D(L - I)x = \mu x \tag{6.31}$$

where μ is an eigenvalue and in general is a complex number and x is a corresponding eigenvector with complex numbers as entries, in general. Without any loss of generality, we assume that x is scaled so that $\|x\| = 1$. Since D is a nonsingular matrix by construction, that is, its inverse exists, we can multiply from the left Eq. (6.31) by D^{-1} and rearrange the terms to get the following:

$$Lx = (\mu D^{-1} + I)x \tag{6.32}$$

and from this expression we derive the following:

$$Lx = (\mu D^{-1} + I)x \Rightarrow 1 > \|L\|\|x\| \geq \|Lx\| = \|(\mu D^{-1} + I)x\|. \tag{6.33}$$

Furthermore,

$$\begin{aligned} 1 > \|(\mu D^{-1} + I)x\| &= \left\| \left[\left(\frac{\mu}{d_1} + 1 \right) x_1, \ldots, \left(\frac{\mu}{d_n} + 1 \right) x_n \right]^T \right\| \\ &\geq \left| \frac{\mu}{d_m} + 1 \right| \|x\| \geq \left| \frac{\mu}{d_m} + 1 \right|, \end{aligned} \tag{6.34}$$

since x is normalized, L satisfies $\|L\| < 1$, and $0 < d_m < 1$ is an entry of D that produces the minimal value of

$$\left| \frac{\mu}{d_m} + 1 \right|$$

over all $i \in \{1, \ldots, n\}$ (may not be unique but only existence is important). Now, let us denote the eigenvalues of J as λ, implying

$$\lambda = \mu + 1,$$

so that from inequality (6.34) one obtains

$$\left| \frac{\lambda - 1}{d_m} + 1 \right| < 1 \Rightarrow |\lambda - (1 - d_m)| < d_m. \tag{6.35}$$

Restriction $0 < d_m < 1$ implies that each eigenvalue λ of J has to be within the interior of a circle centered at $(1 - d_m, 0)$ with radius d_m. This circle is contained within the unit circle and thus the proof. □

A consequence and an interpretation of this result, as presented in Lemma 6.1, is that we can impose well defined restrictions on $g(\cdot)$ so that the overall system is locally stable around an equilibrium. This can be concisely formulated in the following corollary, which again follows from Lemma 6.1.

Corollary 6.1 *If any vector induced matrix norm of the Jacobian of $g(\cdot)$ is strictly smaller than one, then the corresponding linearization of the nonlinear model in Eq.* (6.4) is Schur stable.

Some applications of Corollary 6.1 could be in the cases of "one" and infinity matrix norms that are easy to verify numerically. The conditions are nevertheless slightly more restrictive than requiring the Jacobian of $g(\cdot)$ to be stable but again especially for "one" and the infinity norms the sufficient conditions can be efficiently checked.

6.3 Stability by Linearization of the Gated Recurrent Unit Neural Network Models

In this section, we will concentrate specifically on the gated recurrent unit neural network models and investigate further their special properties when linearized. Again, the main issue with the computation of the linearized model, is the computation of equilibria that can be done easily only when an equilibrium is at zero [4, 5, 22]. Otherwise the equilibria have to be computed numerically.

Now, let us start technical derivations by noting that from Eqs. (6.1) and (6.3) we get the following:

$$g(h) = \tanh(W_g(\varphi(W_o h + b_o) \odot h) + b_g) \tag{6.36}$$

or element wise

$$g_i(h) = \tanh(W_{gi}(\varphi(W_o h + b_o) \odot h) + b_{gi}) \tag{6.37}$$

where, using by now our standard notation, W_{gi} is a row vector denoting the i-th row of W_g and b_{gi} is a scalar denoting the i-th element of b_g. To streamline the presentation, let us introduce a new variable

$$c(h) = \varphi(W_o h + b_o) \odot h. \tag{6.38}$$

In terms of the new variable c, we can compute the following:

$$\frac{\partial g_i}{\partial h} = (1 - g_i^2(h))W_{gi}\frac{\partial c}{\partial h}, \ \forall i \in \{1, \dots, n\}, \tag{6.39}$$

which, in a compact form, can be written as

$$\frac{\partial g}{\partial h} = D_g W_g \frac{\partial c}{\partial h}, \ D_g = \mathrm{diag}\{1 - g_1^2, \dots, 1 - g_n^2\}. \tag{6.40}$$

One of the key ingredients of computing the partial derivative of $g(\cdot)$ is the computation of $\frac{\partial c}{\partial h}$, which follows from the expression

$$c_i(h) = \varphi(W_{oi} h + b_{oi})h_i,$$

where, again following our standard notation, W_{oi} is a row vector denoting the i-th row of W_o and b_{oi} is a scalar denoting the i-th entry of b_o. Now, we can proceed with the computation of the gradient of $c(\cdot)$ as

$$\frac{\partial c_i}{\partial h} = \varphi(W_{oi} h + b_{oi})(1 - \varphi(W_{oi} h + b_{oi}))h_i W_{oi} + \varphi(W_{oi} h + b_{oi})e_i^T. \tag{6.41}$$

Again, the computation of the equilibria is a complex task that, in general has to be done numerically. But if we set $b_g = 0$ then from Eq. (6.3) it follows that there is an equilibrium at zero. Furthermore, substituting $h = b_g = 0$ in Eqs. (6.40) and (6.41) one obtains

$$\left.\frac{\partial g}{\partial h}\right|_{h=0} = W_g D_o, \text{ where } D_o = \operatorname{diag}\{\varphi(b_{o1}), \ldots, \varphi(b_{on})\}. \tag{6.42}$$

If we substitute this expression in Eq. (6.25) and again, for illustration purposes set all the bias vectors to zero, then we obtain the following:

$$\left.\frac{\partial g}{\partial h}\right|_{h=0} = 0.5 W_g. \tag{6.43}$$

It is relevant to note that this computation is consistent with the results reported in [12].

Since the infinity matrix norm is the maximal sum of the absolute values among the rows, it may be the most convenient norm to be applied to the Jacobian expressions in Eqs. (6.40) and (6.41). The result is a bound on the Jacobian of $g(\cdot)$, which can be computed as follows:

$$\left\| \left.\frac{\partial g}{\partial h}\right|_{h=h_e} \right\|_\infty \leq \|W_g\|_\infty (0.25\|W_o\|_\infty + \varphi(\|W_o\|_\infty + b_o^{\max})), \tag{6.44}$$

where

$$b_o^{\max} = \max_{i\in\{1,\ldots,n\}} |b_{oi}|. \tag{6.45}$$

This derivation can be utilized to formulate a stability result in the following theorem.

Theorem 6.2 *If*

$$\|W_g\|_\infty (0.25\|W_o\|_\infty + \varphi(\|W_o\|_\infty + b_o^{\max})) < 1, \ b_o^{\max} = \max_{i\in\{1,\ldots,n\}} |b_{oi}|, \tag{6.46}$$

then the Jacobian evaluated at any equilibrium of the nonlinear model in Eqs. (6.4) and (6.3), is Schur stable, and thus the system is locally Lyapunov stable around each equilibrium.

Proof The proof simply follows Lemma 6.1, its Corollary 6.1, and inequalities (6.44) and (6.46). □

Again, the result formulated in Theorem 6.2 is valid for any equilibrium of the system and it is a consequence that each equilibrium must be inside the unit box $[-1, 1]^n$, that is, $\|D_g\|_\infty < 1$. If appropriate this can also be used in checking the stability properties of the gated recurrent unit neural network models as formulated in the following remark.

Remark 6.1 The inequality (6.44) can be tested for a trained gated recurrent unit neural network, that is, when all the entries of the weight matrices and bias vectors are computed and the test requires no equilibria computation.

6.3.1 Learning with Gated Recurrent Unit Neural Networks Without the Reset Gate

In this subsection, we will consider a special type of learning that can be related to the equilibria and stability analysis. Since in this setup, the equilibria will be connected to the data, in terms of learning it is more suitable to think about image or symbol classifications rather then time-varying sequences. To have this connection between learning and equilibria analysis established in a meaningful way we will consider a special version of the gated recurrent unit neural network being the one without the reset or output gate. The dynamical discrete-time model for this type of recurrent neural networks is a set of difference equations represented as follows [16]:

$$\begin{aligned} h(k+1) = \; & \varphi_f(h(k), r(k)) \odot h(k) + (\mathbf{1} - \varphi_f(h(k), r(k))) \\ & \odot \tanh(W_g h(k) + V_g r(k) + b_g) \equiv G(h(k), r(k)), \\ \text{where } \varphi_f(h(k)) = \; & \varphi(W_f h(k) + V_f r(k) + b_f)). \end{aligned} \tag{6.47}$$

Following our notation, $h(\cdot) = [h_1(\cdot), \ldots, h_n(\cdot)]^T$ denotes the n-dimensional both state and output vector and $r(\cdot) = [r_1(\cdot), \ldots, r_n(\cdot)]^T$ denotes either the training signal or the signal to be classified. Furthermore, from the control theory point of view, the signal to be classified can be considered as the reference or input signal. So in this model we do include the input signal. This renders the model to be time-varying if the reference signal is time-varying and time-invariant if the reference signal is constant. The inclusion of the reference signal requires the incorporation of additional weight matrices V_g and V_f. The simplification due to removing the reset gate results in

$$\varphi_o(h) \odot h \equiv h.$$

Now, having introduced the network dynamics, let us describe the setup of the learning process. Because this is an equilibrium based training, we consider time-invariant training signals, that is,

$$r(k) \equiv \bar{r}, \tag{6.48}$$

which is a scenario that, again, resembles image classification, and choose a simple task that the output “learns" the input signal. Keep in mind that in this specific case the output is the state vector. Mathematically this can be expressed as

$$\lim_{k \to \infty} h(k) = \bar{r} \tag{6.49}$$

and furthermore implies

$$g(h) = \tanh(W_g h + V_g \bar{r} + b_g), \quad \text{with} \quad \frac{\partial g}{\partial h} = D_g W_g, \tag{6.50}$$

since in this case

$$c(h) \equiv h.$$

If the total number of constant training signals is denoted by P then the objective of the learning process is that the output converges to the corresponding input, that is,

$$\lim_{k\to\infty} h_p(k) = \bar{r}_p, \ \forall p \in \{1, \ldots, P\}. \tag{6.51}$$

Therefore, the process of learning and training is expected to depend on two components. One is the convergence, if it happens and if it does how fast will it be and the other is how well does the output signal match the input signal. By matching we mean if the convergence happens would it be perfect as indicated in Eq. (6.51) or would there be an error. The convergence should at least locally depend on the Jacobian and its eigenvalues where the corresponding equilibria are solutions to the following algebraic equations:

$$(W_g + V_g)\bar{r}_p + b_g = \tanh^{-1}(\bar{r}_p) = q_p, \ \forall p \in \{1, \ldots, P\}. \tag{6.52}$$

Transposing each individual equation in Eq. (6.52) results in

$$[\bar{r}_p^T, 1]\begin{bmatrix} W_g^T + V_g^T \\ b_g^T \end{bmatrix} = q_p^T, \ \forall p \in \{1, \ldots, P\}, \tag{6.53}$$

and concatenating all equations into one, results in the following expression:

$$R\begin{bmatrix} W_g^T + V_g^T \\ b_g^T \end{bmatrix} = Q \Rightarrow \begin{bmatrix} W_g^T + V_g^T \\ b_g^T \end{bmatrix} = R^\dagger Q, \tag{6.54}$$

where R is the matrix constructed by stacking P rows of the form $[\bar{r}_p^T, 1]$, and Q is the matrix constructed by stacking P rows of the form q_p^T. Therefore, R is an $P \times (n+1)$-dimensional matrix and Q is an $P \times n$-dimensional matrix. If R has full row rank, its right pseudo-inverse can be computed as

$$R^\dagger = R^T(RR^T)^{-1} \tag{6.55}$$

and if not then a generalized pseudo-inverse [6], denoted also as $R^\dagger$, should be computed. Notice that the set of equations in (6.52) represent a set of algebraic equations which may not be even feasible but we can at least minimize the deviation by computing pseudo-inverses. This method is essentially the projection learning rule (for more details see, for example, Sect. 8.2 in [15]) which has been used more than two decades ago for classification of constant signals like symbols or images. An important difference is that the neural networks used then had simpler dynamics to analyze. Furthermore, the same formulation can be used when the input and output signals do not match, for example, when different output is used to classify the input signal, which will result in a different matrix entries on the righthand side of Eq. (6.54).

The other component is to establish convergence and the convergence characteristics, if possible. The convergence can be considered globally or locally. As for the characteristics of the convergence, if it does happen, one of the important ones is how fast the convergence is. In the local case, the speed of convergence is related to the eigenvalues of the corresponding Jacobian. If the eigenvalues are closer to the origin then the convergence is faster, so that the fastest convergence is the one when all of the eigenvalues are set to zero. Now, the question is if this is desirable for other reasons where one important observation is, to mention once again, that the trained recurrent neural networks which perform well, are observed to operate at the "edge of chaos" [2]. But the edge of chaos has been observed to be closely related to the boundary stable equilibria. These seems to be contradicting requirements yet we will not get into any deeper discussion how to resolve the issue, at least for now.

By computing the Jacobian for the gated recurrent unit neural network model without the reset gate one obtains the following compact form:

$$J = \left. \frac{\partial G(h)}{\partial h} \right|_{h=h_e=\bar{r}} = D_1 + (I - D_1) D_g W_g . \tag{6.56}$$

In Eq. (6.56), one of the components being D_1, as defined in Eq. (6.25), depends on W_f, V_f and b_f, which do not influence the equilibria fitting Eq. (6.54). Therefore, technically speaking these matrices and the bias vector can be used to control the placement of the eigenvalues of the Jacobians. Multiple Jacobians will exist due to the general Jacobian form being dependent on the equilibria, that is, the training signals. One possible way of doing this control or tuning of the eigenvalues, is to simply set D_1 as δI, where $\delta \in (0, 1)$. Notice that the idea behind this tuning is also "convex combination" alike. This is related to the convex combination formulation because for the smaller values of δ the eigenvalues of the Jacobians would be closer to the eigenvalues of $D_g W_g$ and for the larger values of δ the eigenvalues would get closer to δ and thus to the boundary of stability and therefore the "edge of chaos." Of course the designer may choose different forms for all Jacobians yet that would significantly complexify the design and it is unclear at this point what the benefits would be. Now, if all Jacobians are to be the same, denoted as $\bar{J}$ then from Eq. (6.56) it follows that

$$D_1 = (\bar{J} - D_g W_g)(I - D_g W_g)^{-1}.$$

But this equation, at least form wise, poses some challenges. One is that no eigenvalue of $D_g W_g$ equals one, so the matrix $I - D_g W_g$ is nonsingular, is simple to achieve. The other one, which is that

$$(\bar{J} - D_g W_g)(I - D_g W_g)^{-1}$$

has to be a diagonal matrix with diagonal entries having values between zero and one, is much more difficult to accommodate. One simple way to resolve this issue is to impose on W_g to be diagonal, yet this could be overly restrictive in terms of the network's performance. This discussion is only to illustrate the procedure and any substantial conclusions would depend on further investigations. Regardless of this, once D_1 is set, resulting in D_1^p, corresponding to each training signal $\bar{r}_p$, one obtains the following:

$$(W_{fi} + V_{fi})\bar{r}_p + b_{fi} = \varphi^{-1}(m_{ii}^p) = z_i^p, \; i \in \{1, \ldots, n\}, \; p \in \{1, \ldots, P\}, \tag{6.57}$$

with m_{ii}^p being diagonal entries of matrices D_1^p, $p \in \{1, \ldots, P\}$. The linear equations in (6.57) could now be used for designing W_f, V_f, and b_f. These linear equations can also be written as

$$(W_f + V_f)\bar{r}_p + b_f = z^p, \; \forall p \Rightarrow (W_f + V_f)[\bar{r}_1, \ldots, \bar{r}_P] = C = [z^1 - b_f, \ldots, z^P - b_f] \tag{6.58}$$

with

$$z^p = [z_1^p, \ldots, z_n^p]^T, \; \forall p.$$

In the compact form Eq. (6.58) can be written as

$$R_1(W_f^T + V_f^T) = C^T \Rightarrow W_f + V_f = C R_1^{\dagger T} \tag{6.59}$$

where $R_1 = [\bar{r}_1, \ldots, \bar{r}_P]^T$, and $R_1^\dagger$ represent the pseudo-inverse or the generalized pseudo-inverse of R_1.

To conclude this subsection, let us formalize steps for the suggested learning procedure, as follows. First, we compute b_g and $W_g + V_g$ explicitly from the equilibria inspired Eq. (6.54). Since only the sum $W_g + V_g$ is determined explicitly one can set W_g to be equal to any desired matrix and then compute V_g simply as $S - W_g$ where $S - W_g + V_g$. To establish convergence we can impose $\|W_g\| < 1$ inequality condition or even less restrictive inequality condition $\|W_g\| < 1/\|D_g\|$ (again $0 < \|D_g\| < 1$ is satisfied by construction), where the later follows from Corollary 6.1 and Eq. (6.50). After computing W_g, V_g, and b_g, the next step is to compute W_f, V_f, and b_f to satisfy Eq. (6.59), and accomplish fine tuning of the Jacobian at equilibria. Depending on the consistency of these equations the computation of the pseudo-inverse or, if needed, the generalized pseudo-inverse will lead to, in general, approximate solutions to the equations.

References

1. R. B. Barmish. *New Tools for Robustness of Linear Systems*. Macmilllan Publishing Company, New York, NY, 1994.
2. N. Bertschinger and T. Natschläger. Real-time computation at the edge of chaos in recurrent neural networks. *Neural Computation*, vol. 16, no. 7, pp. 1413–1436, 2004.
3. K. Cho, B. V. Merrienboer, C. Gulcehre, D. Bahdanau, F. Bougares, H. Schwenk, and Y. Bengio. Learning Phrase Representations using RNN Encoder-Decoder for Statistical Machine Translation. arXiv:1406.1078, 2014.

4. S. A. Deka, D. M. Stipanović, B. Murmann, and C. J. Tomlin. Stabilization of long short-term memory neural networks with constant weights and biases. *Journal of Optimization Theory and Applications*, vol. 181, pp. 231–243, 2019.
5. S. A. Deka, D. M. Stipanović, B. Murmann, and C. J. Tomlin. Long-short term memory neural network stability and stabilization using linear matrix inequalities. *Proceedings of the 2019 IEEE International Symposium on Circuits and Systems*, 2019.
6. D. S. Djordjević and V. Rakočević. *Lectures on Generalized Inverses*. Faculty of Sciences and Mathematics, University of Niš, Niš, Serbia, 2008.
7. A. F. Filippov. *Differential Equations with Discontinuous Righthand Sides*. Kluwer Academic Publishers, Dordrecht, The Netherlands, 1988.
8. A. Graves. Generating sequences with recurrent neural networks. *arXiv*:1308.0850, 2013.
9. K. Greff, R. K. Srivastava, J. Koutník, B. R. Steunebrink, and J. Schmidhuber. LSTM: A search space odyssey. *IEEE Transactions on Neural Networks and Learning Systems*, vol. 28, no. 10, pp. 2222–2232, 2017.
10. M. W. Hirsch, S. Smale, and R. L. Devaney. *Differential Equations, Dynamical Systems, and an Introduction to Chaos*. 3rd Edition, Elsevier Inc., Waltham, MA, 2013.
11. S. Hochreiter and J. Schmidhuber. Long Short-Term Memory. *Neural Computation*, vol. 9, no. 8, pp. 1735–1780, 1997.
12. S. Kanai, Y. Fujiwara and S. Iwamura. Preventing gradient explosions in gated recurrent units. *Proceedings of the 2017 Neural Information Processing Systems (NIPS) Conference*, 2017.
13. J. P. LaSalle. *The Stability and Control of Discrete Processes*. Springer-Verlag, New York, NY, 1986,
14. G. A. Leonov. *Strange Attractors and Classical Stability Theory*. St. Petersburg University Press, St. Petersburg, Russia, 2008.
15. A. N. Michel and D. Liu. *Qualitative Analysis and Synthesis of Recurrent Neural Networks*. Marcel Dekker, New York, NY, 2002.
16. M. Ravanelli, P. Brakel, M. Omologo, and Y. Bengio. Light gated recurrent units for speech recognition. *IEEE Transactions on Emerging Topics in Computational Intelligence*, vol. 2, no. 2, pp. 92–102, 2018.
17. D. S. Shevitz and B. E. Paden. Lyapunov stability theory of nonsmooth systems. *IEEE Transactions on Automatic Control*, vol. 39, pp. 1910–1914, 1994.
18. S. S. Stanković, M. S. Stanković and D. M. Stipanović. Decentralized parameter estimation by consensus based stochastic approximation. *IEEE Transactions on Automatic Control*, vol. 56, pp. 531–543, 2011.
19. D. M. Stipanović. *A New Scheme for Controlling Nonlinear Systems*. Diploma Thesis, School of Electrical Engineering, University of Belgrade, Belgrade, Yugoslavia, 1994.
20. D. M. Stipanović and D. D. Šiljak. Stability of polytopic systems via convex M-matrices and parameter-dependent Liapunov functions. *Nonlinear Analysis*, vol. 40, pp. 589–609, 2000.
21. D. M. Stipanović and D. D. Šiljak. Connective stability of discontinuous dynamic systems. *Journal of Optimization Theory and Applications*, vol. 115, pp. 711–726, 2002.
22. D. M. Stipanović, B. Murmann, M. Causo, A. Lekić, V. R. Royo, C. J. Tomlin, E. Beigne, S. Thuries, M. Zarudniev, and S. Lesecq. Some local stability properties of an autonomous long short-term memory neural network model. *Proceedings of the 2018 IEEE International Symposium on Circuits and Systems*, 2018.
23. D. M. Stipanović, M. N. Kapetina, M. R. Rapaić, and B. Murmann. Stability of Gated Recurrent Unit Neural Networks: Convex Combination Formulation Approach. *Journal of Optimization Theory and Applications*, vol. 188, pp. 291–306, 2021.

Part III

Modified Internal Model Control and Recurrent Neural Networks

7 A Modified Internal Model Control Approach

One of the most, and at the same time, relevant and challenging tasks when dealing with dynamical systems is the problem of tracking, which can also be viewed as a problem of regulation of dynamical systems in control theory. From our prospective, this problem is not only relevant as is but, in our opinion, can be related to neural networks and their application in machine learning. In particular, the problem of classification of input signals as elements of time-varying sequence or image datasets may be considered as a problem of how well the system tracks the input signal, that is, how well the input signal can be recognized or classified at the system's output. The rational behind this is that the reference input signal would be tracked or followed and thus classified easier and more accurately if the system is capable of recognizing it. Among process control schemes, the internal model control [4–6, 13, 14] is chosen for its simplicity of formulation and reliable and accurate performance for a large scale of complex dynamical systems. By performance we mean that the system tracks a wide variety of, in general, time-varying reference signals within the desired accuracy specifications. Depending on the information about the process, a number of different identification procedures can be performed to model the process. If the process is treated as a "black box" then usually identification is done based on a class of training signals. This class of training signals has to be carefully chosen so that the system would be able to track both signals which belong to the same class and the ones that don't. Thus, not only it is important to be able to learn about the process based on the input-output relationship but this learning will also depend on the choice of the reference signals for which this relationship is established. For example, a class of smoothen step signals was found to be very effective in identifying complex and, for the designer, unknown nonlinear systems [19]. This may be explained by recognizing that a smoothen step signal while simple has wider spectrum than a constant or a sinusoidal signal which, on their own, contain only one

D. Stipanović, *Difference Equations and Machine Learning*, Synthesis Lectures on Mathematics & Statistics, https://doi.org/10.1007/978-3-032-00910-4_7

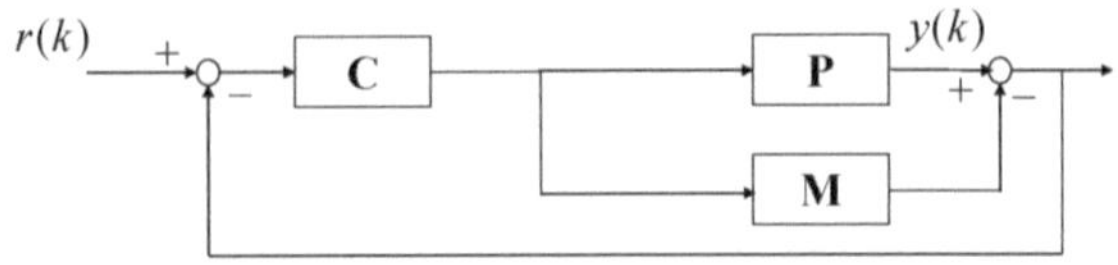

Fig. 7.1 Internal model control block diagram

component in the spectrum. On the other hand, how to choose training signals in machine learning applications may be restricted by the available datasets and their contents.

To introduce the internal model control approach to control and regulate dynamical systems, a basic internal model control block diagram is depicted in Fig. 7.1.

This block diagram does not only depict internal model control scheme but also a philosophy behind a specific control of a process considered as a dynamical system. In this block diagram, **P** denotes a process to be controlled, its model is denoted by **M**, and **C** denotes a controller. In general, the time-varying input reference signal is denoted as $r(k)$ and the goal is to enable the output of the system denoted by $y(k)$ to track the reference signal. The performance of the internal model control approach is measured by how well the reference signal is tracked by the system's output. Thus, this approach belongs to the class of regulation control approaches specifically designed to track the reference signal.

Let us first discuss the controller block within the internal model block diagram. This controller block is most often considered to be a combination of a filter and an inverse model. The filter is usually chosen to smoothen the transient behaviors of the signals and therefore most commonly it represents a low-pass filter which removes faster components in the spectrum. In the standard feedback block diagrams an integrator is placed at this position to enhance the tracking yet withing the internal model control schematic this is not needed due to the feedforward branch. The filter is denoted as **F** and the inverse model as **IM**, as illustrated in Fig. 7.2.

One simple observation is that the internal model control approach to the regulator problem does involve not only identifying the model but also the inverse model of the process. This is not an easy task and may include a number of issues. For example, if one of the two is slower and easier to identify then the other may be quite difficult to handle, given that these two blocks are inverses of each other. An interpretation for this would be a scalar function which does not change much in terms of its image values, which as a consequence implies a very steep inverse function. Motivated by trying to overcome this issue, we will illustrate the process that led not only to a simplification, but to a more effective regulation or tracking block diagram based on the same principle. Furthermore, the effectiveness and advantages

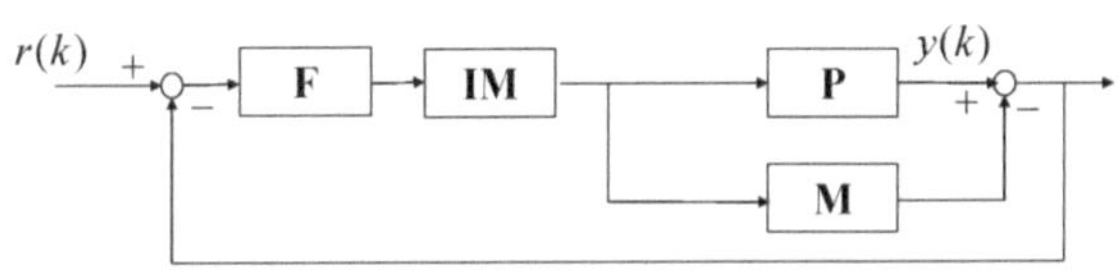

Fig. 7.2 An internal model control block diagram with a filter and an inverse model

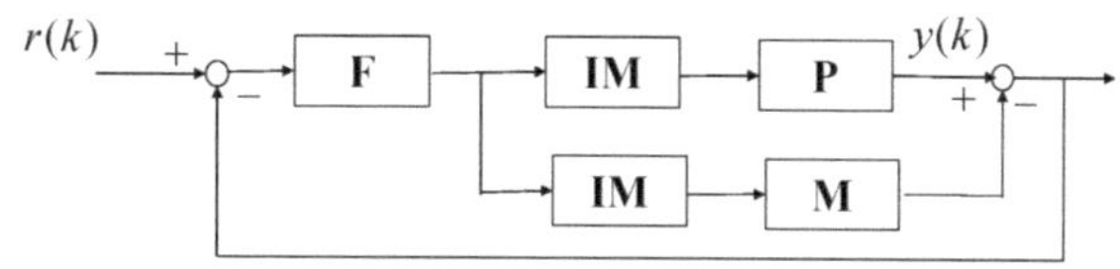

Fig. 7.3 A reconfigured basic internal model control block diagram

of the simplification, as it will be argued, will be explained and justified in technical terms, to a certain extent. The process of simplification starts by pushing the inverse model block forward, which is an input/output equivalent transformation, that is, renders the input/output relationship invariant. This simple transformation results in a new block diagram depicted in Fig. 7.3.

Now, the serial combination of the two blocks in the feedforward branch, that is, the inverse model and the model itself can be replaced by the identity. Notice that this, interesting enough, is process independent and trivial given that the inverse model has to be an inverse of the model not the process. This is obvious and a consequence of both of these blocks being created by the designer and thus replacing them with an identity is trivially justified [19]. Now, from a practical point of view, a delay can and should be inserted if for no other reason then to match the time the signals travel while processed within the main branch. We will hold on doing this explicitly for a moment so that we can provide a simple derivation of an intuitive tracking capabilities of the block diagram with no delay. This will provide a simple, yet very illustrative, technical insight into why the modified block diagram has the capability of the perfect tracking performance. By doing so, that is, replacing these two blocks with an identity block manipulation leads to a simplification of the internal model block diagram provided in Fig. 7.4. We will refer to this block diagram as a basic modified internal model control block diagram. Again, the modified block diagram with no delay in the feedforward branch is valuable since it is very convenient to illustrate the inherently perfect tracking of the modified internal model control scheme.

To provide a simple yet effective and illustrative technical arguments to support the modification and simplification of the internal model control block diagram let us even remove the filter block from the block diagram. By removing the filter, we get even simpler block diagram, which is provided in Fig. 7.5. In this block diagram, the error signal, which is a discrepancy between the output and the reference signal, is depicted and denoted as $e(k)$ to enable a technical justification of the suggested perfect tracking capabilities of the modified scheme.

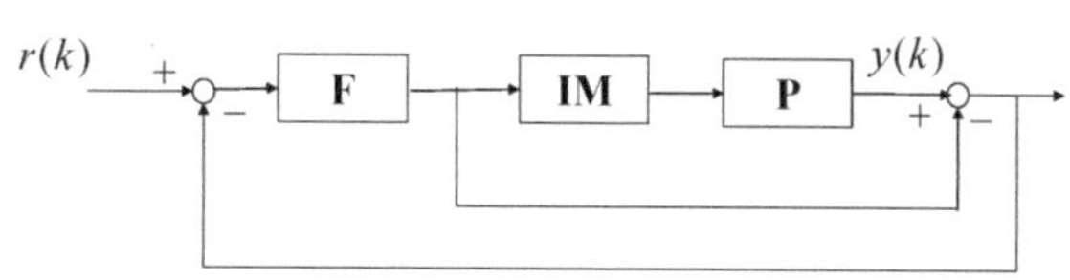

Fig. 7.4 Basic modified internal model control block diagram

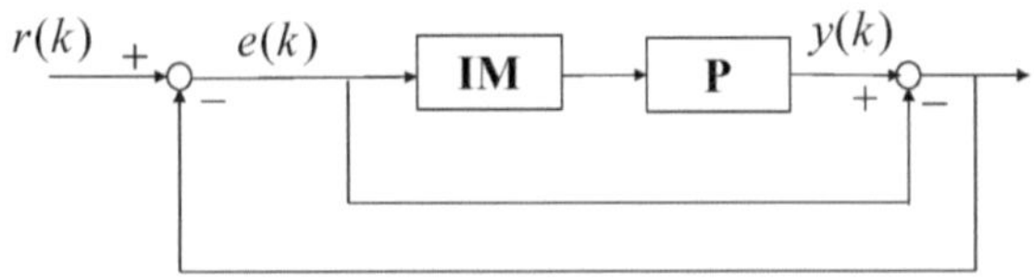

Fig. 7.5 Simplified modified internal model control block diagram without a filter

Now, we aim to provide a simple technical justification of the tracking properties of the basic modified internal model control block diagram, which is mainly due to the feedforward branch. First thing to notice is that the gain of the feedforward branch equals one, which could be interpreted as a generalized integrator action in basic terms, thus also suited for the nonlinear systems. Related to simplified model depicted in Fig. 7.5, obtained after removing the filter block, we can write the signal equation for the point in the block diagram which is just after the initial summation point on the left, as follows:

$$e(k) = r(k) - (y(k) - e(k)) \Rightarrow y(k) = r(k). \tag{7.1}$$

This simple Eq. (7.1) is essentially a trivial technical observation [19], which is derived from simplified and, in a way, idealized modified internal model control block diagram. This simplification serves the purpose of boldly illustrating the idea of the possibility of perfect tracking and thus provides the motivation behind the simplification of the internal model control scheme. This scheme not only does not require an inverse model but it also does not require even a model of the process. Therefore, in this idealistic scenario, a perfect tracking is established for any process and with simple implementation that requires neither an inverse model nor a model of the process. Of course we have to keep in mind that this is only an idealized scenario that cannot be applied in this idealized form in real-time control of processes.

Given that the basic modified internal model control scheme itself enables the system to track the reference input perfectly, if feasible, we can replace the inverse model block with a controller, without essentially jeopardizing the tracking capability of the scheme. By feasibility we mean that there exists an input to the process that would produce the reference signal as the output because otherwise it would not be physically possible for the process to follow the reference input. This is an essential and basic assumption that a given process or system is capable to track the reference signal, and in that case the process needs to be controlled and enabled to do so. The control is provided by the tracking scheme and its controller blocks. Now, if the feasibility is not guaranteed, then it would be desirable to at least guarantee boundedness of all signals within the tracking scheme and within the blocks, which we will refer to as internal stability. Thus, the design of the controller in this context would need to, if feasible, enable tracking, that is, that the output of the controller, as the input to the system, would result in the tracking signal as the output of the system, while keeping all signals bounded and well behaved. This inherently means that an internal stability and feasibility of the system would be both established and the block diagram would

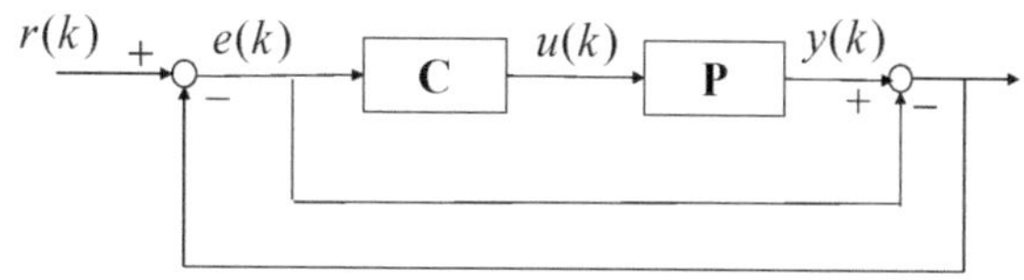

Fig. 7.6 Simple modified internal model control block diagram with a controller instead of an inverse model

properly work both from the input/output as well as from the internal operation prospective. This would also mean that the regulation task is fully accomplished and would be performed safely and reliably. The corresponding block diagram is depicted in Fig. 7.6, and the scheme is another simple version of a modified internal model control block diagram [19]. Notice again, that in this scheme the controller does not contribute to tracking and its sole purpose would be to establish, again, internal stability and feasibility.

Stating that in the modified internal model control block diagram the controller only needs to stabilize the system and provide a feasible input to the process, so that the output matches the desired reference signal without being disrupted by the internal instability and infeasibility, is a bit incomplete since it may not be as easy as it seems, in general. An appropriate stabilization concept, which would provide boundedness of the internal signals, for nonlinear systems, is the concept of internal stability and the small gain stability approach based on the input to state and output stability results for discrete-time systems [7, 9, 10]. Establishing these conditions would provide proper functionality of the corresponding block diagram in terms of boundedness of all of its internal and external signals.

The perfect tracking is, as mentioned earlier, not realizable in real-time due to the time that takes signals to pass, at least, through the process block if not in addition though the other blocks in the main branch. Therefore, if there is a controller block preceding the process, the time to be accommodated for must include both blocks. This is why the feedforward branch needs to include a delay to accommodate for the processing time of signals in the main branch, and thus for the scheme to be implementable in real-time. The process itself is of dynamic nature, and thus incurs at least one time step of delay. The same minimal time delay of one time instance for the controller must be assumed too. Therefore, at least a few time steps are spent for processing signals in the main branch including the process and its inverse model or a controller. In our earlier investigations, it has been shown in simulations that it would be beneficial to add a time delay in the corresponding feedforward branch which approximated the delay in the main branch. The modified internal model control block diagram with a delay in the feedforward branch is shown in Fig. 7.7. The earlier works were published originally in [19] and later in [11, 12], where the list of authors is in alphabetical order. Time delay block is denoted by D where the actual duration of the delay is up to the designer to properly estimate, depending on the process itself.

Therefore, to satisfy the basic implementation requirement we need to include a delay in the feedforward branch as well as a filter in the main branch for a more general and more realistic version of the modified internal model control block diagram. This, more sophisticated and appropriate to implement block diagram is depicted in Fig. 7.8. In a comprehensive review of modified internal model control structures [16] neither the structure in Fig. 7.7 nor the structure in Fig. 7.8 were mentioned. A slightly modified structure as the one in Fig. 7.7, where the filter is not included, has been derived and mentioned in [18], yet it has been disregarded because of the reported performance that was not considered satisfactory. This reported experience confirms that it is better, performance and operation wise, to integrate a general controller that would enable feasibility and stability of the modified internal model control block diagram rather than an inverse model of the process. This is in line with what has been mentioned earlier that there is no particular need of having an inverse model preceding the process. Therefore, it is much more general and beneficial in terms of the overall regulation control performance to assume that the inverse block is replaced by a controller. In the block diagram we also indicate $q(k)$ as a time-varying variable that will be used in a signal derivation to analyze the modified internal model control block diagram.

At this point, let us assume that the dynamics of the filter block is represented by a very general form, that is, using the following form of nonlinear discrete-time equations that consist of both difference state-space variable and algebraic output equations:

$$\begin{aligned} x_f(k+1) &= f_f(x_f(k), e(k)), \\ q(k) &= g_f(x_f(k), e(k)), \end{aligned} \tag{7.2}$$

where x_f denotes the filter's vector of state variables and q is its output. Given that the controller is another dynamical system, it can be represented in the same general form as follows:

$$\begin{aligned} x_c(k+1) &= f_c(x_c(k), q(k)), \\ u(k) &= g_c(x_c(k), q(k)), \end{aligned} \tag{7.3}$$

where x_c denotes the controller's state vector, and u denotes its output, which is also the input to the process.

Finally, the process is another dynamical system, which is also assumed to be in a general form including nonlinear state difference and output algebraic equations. Therefore, we model dynamics of the process using the following representation:

$$\begin{aligned} x(k+1) &= f(k, x(k), u(k)), \\ y(k) &= g(k, x(k), u(k)), \end{aligned} \tag{7.4}$$

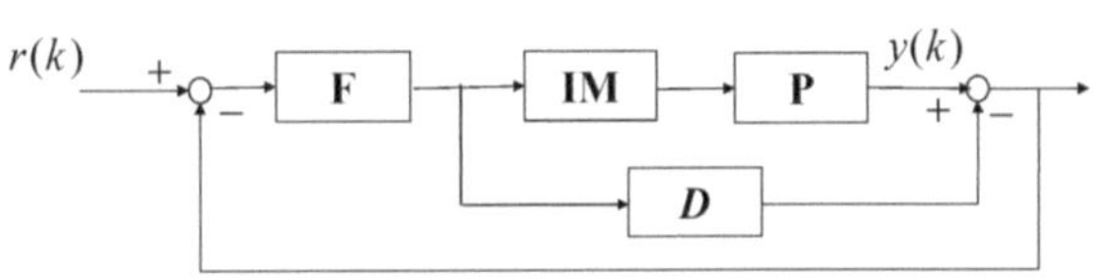

Fig. 7.7 Modified internal model control block diagram with a delay

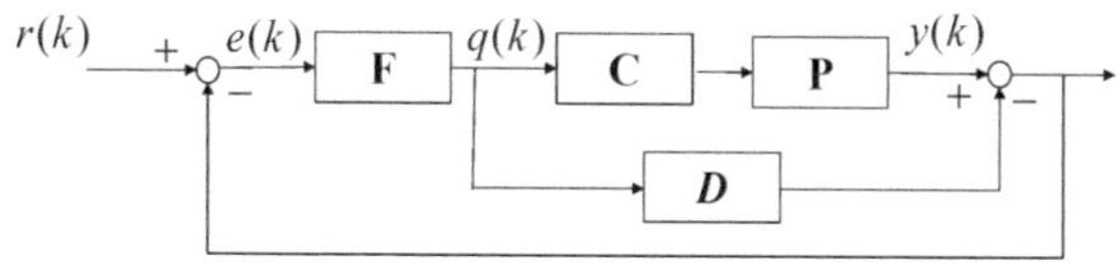

Fig. 7.8 Modified internal model control block diagram with a filter and delay

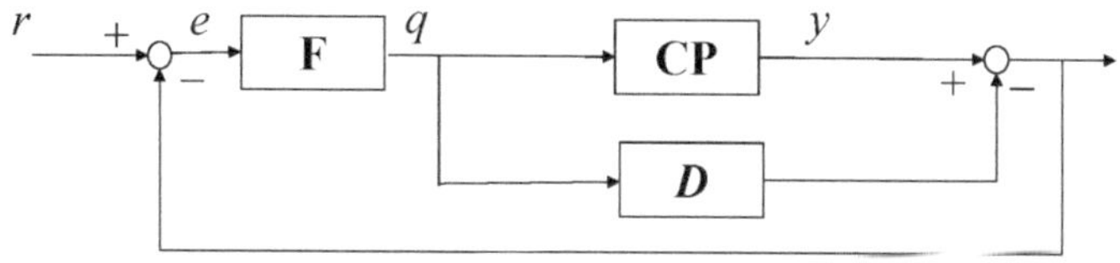

Fig. 7.9 Modified internal model control block diagram with a combined block for the process and the controller

where x is the state of the process. Given that the process is not being created by the designer, but given to the designer to be controlled, we assume also that it may be time-varying, in general.

The block diagram difference equations can be put in an even more compact form if we assume that we can combine dynamics of the controller and the process as one dynamical system. Therefore, to shorten the notation, we assume that the representation of the serial connection of the controller and the process is in the form of one dynamic system. Given that the process is time-varying we assume the same for the concatenation and arrive at the following:

$$y(k+d) = g(k, y(k+d-1), \ldots, y(k), q(k)), \tag{7.5}$$

where the duration of the delay is approximated by d sample periods of time. Let us now simplify the presentation even more by introducing a mildly conservative assumption that we do not need the output equation for the filter, that is, that the output of the filter equals to the state vector of the filter. Then, the overall dynamics can be represented as follows:

$$\begin{aligned} y(k+d) &= g(k, y(k+d-1), \ldots, y(k), q(k)) \\ q(k+1) &= f_f(q(k), e(k)) \\ e(k) &= r(k) - y(k) + q(k-d) \end{aligned} \tag{7.6}$$

This new set of formulations is illustrated in Fig. 7.9. Furthermore, we drop the time-dependence notation on the block diagram to pay more attention to signals themselves. The controller and the process blocks are combined into one block simply denoted as **CP**.

By substituting the equation for the error signal into the filter dynamical model in Eq. (7.6), we get the following slightly more concatenated system of difference equations:

$$\begin{aligned} y(k+d) &= g(k, y(k+d-1), \ldots, y(k), q(k)) \\ q(k+1) &= f(q(k), r(k) - y(k) + q(k-d)) \end{aligned} \tag{7.7}$$

We will not be concentrating on the modified internal model control stability properties perse in this book. Instead, we will rather concentrate on a particular realization using neural networks. Earlier, either the standard or modified internal model control block diagrams, have been used for training and implementation of neural networks [8, 11, 12, 15, 17–19]. More recently, gated recurrent unit neural networks [2] were used to build the model and the inverse model blocks to synthesize the internal model regulation for stable nonlinear systems [1]. This chapter serves the purpose of introducing the reader to the standard internal model control scheme and then provide a motivation behind using a modified and easier to implement version of it, to which we refer as the modified internal model control scheme.

In this monograph, our idea to use modified internal model control scheme is to enhance the performance of the classification of a given neural network. Basically, the idea is to use an already trained neural network and to replace the process itself with this neural network. Then, the modified internal model scheme itself will enable the neural network to better and more reliably track the input signal, and thus will improve the process of classification. By assuming that the process is now a neural network with sigmoidal outputs, we will exploit the fact that the outputs of the process are bounded by construction and provide some technical insights into the tracking performance of this specific modified internal model control setup. This will be covered in the material presented in the next chapter.

References

1. F. Bonassi and R. Scattolini. Recurrent Neural Network-based Internal Model Control design for stable nonlinear systems. *European Journal of Control*, vol. 65, 100632, ISSN 0947-3580, 2022.
2. K. Cho, B. V. Merrienboer, C. Gulcehre, D. Bahdanau, F. Bougares, H. Schwenk, and Y. Bengio. Learning Phrase Representations using RNN Encoder-Decoder for Statistical Machine Translation. arXiv:1406.1078, 2014.
3. A. Datta. *Adaptive Internal Model Control*. Springer-Verlag, London, 1998.
4. C. G. Economou, M. Morari and B. O. Palsson. Internal Model Control: Extension to Nonlinear Systems. *Industrial & Engineering Chemistry Process Design and Development*, vol. 25, no. 2, 403–411, 1986.
5. C. E. Garcia and M. Morari. Internal Model Control: A Unifying Review and Some New Results. *Industrial & Engineering Chemistry Process Design and Development*, vol. 21, no. 2, 308–323, 1982.
6. C. E. Garcia and M. Morari. Internal Model Control: Multivariable Control Law Computation and Tuning Guidelines. *Industrial & Engineering Chemistry Process Design and Development*, vol. 24, 484–494, 1985.
7. R. Geiselhart and F. R. Wirth. Relaxed ISS Small-Gain Theorems for Discrete-Time Systems. *SIAM Journal on Control and Optimization*, vol. 54, pp. 423–449, 2016.
8. K. J. Hunt and Sbarbaro. Neural Networks for Nonlinear Internal Model Control. *IEE Proceedings D of Control Theory and Applications*, vol. 138, pp. 431–438, 1991.

9. Z. P. Jiang and Y. Wang. Input-to-state stability for discrete-time nonlinear systems. *Automatica*, vol. 37, pp. 857–869, 2001.
10. Z. P. Jiang, Y. Lin, and Y. Wang. Nonlinear small-gain theorems for discrete-time feedback systems and applications. *Automatica*, vol. 40, pp. 2129–2136, 2004.
11. M. R. Mataušek and D. M. Stipanović. A New Approach to Nonlinear Control System Design. *Journal of Automatic Control*, vol. 5, pp. 31–42, 1994.
12. M. R. Mataušek and D. M. Stipanović. Modified Nonlinear Internal Model Control. *Control and Intelligent Systems*, renamed to *Mechatronic Systems and Control*, vol. 26, pp. 57–63, 1998.
13. M. Morari. Internal Model Control - Theory and Applications. *IFAC Proceedings Volumes*, vol. 16, pp. 1–18, 1983.
14. M. Morari and E. Zafiriou. *Robust Process Control*. Prentice Hall, Englewood Cliffs, NJ, 1989.
15. E. P. Nahas, M. A. Henson and D. E. Seborg. Nonlinear Internal Model Control Strategy for Neural Network Models. *Computers & Chemical Engineering*, vol. 16, no. 12, 1039–1057, 1992.
16. A. Ranjan, U. Mehta and S. Saxena. A comprehensive review of modified Internal Model Control (IMC) structures and their filters for unstable processes. *Annual Reviews in Control*, vol. 56, pp. 1–14, 2023.
17. I. Rivals and L. Personnaz. Nonlinear Internal Model Control Using Neural Networks: Application to Processes with Delay and Design Issues. *IEEE Transactions on Neural Networks*, vol. 11, no. 1, pp. 80–90, 2000.
18. D. C. Psichogios and L. H. Ungar. Direct and Indirect Model-Based Control Using Artificial Neural Networks. *Industrial and Engineering Chemistry Research*, vol. 30, no. 12, pp. 2564–2573, 1991.
19. D. M. Stipanović. Controlling Nonlinear Systems using Neural Networks. *Diploma Thesis*, Department of Automatic Control, School of Electrical Engineering, University of Belgrade, 1994.

8 Modified Internal Model Control and Neural Networks

How to efficiently and effectively integrate feedback information in controlling and regulating dynamical systems is an important yet very challenging and complex task. The complexity stems not only from the complexity of dynamical systems [11, 17] but also from the quality and accuracy of the feedback information. Given its well-known performance in enabling and regulating dynamical systems to track reference signals, we plan to explore the use of the modified internal model control approach, where the basics were also presented in the previous chapter. While the internal model control approach [5–8, 21, 22] can be used both in training and the implementation of neural networks [3, 13, 19, 20, 24], we will concentrate on the implementation part, but in a nonstandard way as it will be shown. In line of this, we will assume that the recurrent neural network is of generic type, and has already been trained, without getting into too much specific details on how it has been trained. Some dynamical properties of the long short-term memory and gated recurrent unit neural networks have been discussed in previous chapters of this book, yet the presentation in this chapter will be kept independent and applicable to general recurrent neural networks. Furthermore, by formulating our approach in a specific way we will be able to bypass dynamical intricacies and complexities that some specific recurrent neural networks like long short-term memory neural networks impose yet still be able to integrate them into our formulation. This specific formulation and integration will be based on our motivation to use modified internal model control approach to enhance classification capabilities of a given recurrent neural network to be regulated to track and recognize input reference signals. As mentioned in the previous chapter, the main motivation for the tracking specification and formulation with recurrent neural networks is that the problem of tracking an unknown reference signal is essentially related to recognizing or classifying that time-varying or time-invariant reference input signal.

D. Stipanović, *Difference Equations and Machine Learning*, Synthesis Lectures on Mathematics & Statistics, https://doi.org/10.1007/978-3-032-00910-4_8

8.1 Modified Internal Model Control Neural Network Implementation

A version of the basic internal model control scheme, customized toward the integration of recurrent neural networks, is depicted in Fig. 8.1. This block diagram incorporates the idea of how a general process is being controlled to track the reference signal, using the basic internal model control scheme and recurrent neural networks. In this setup, the internal model control block diagram consists of a process **P**, model of the process designed using neural networks **PNN** and the inverse model also designed using neural networks **INN**. Again, this scheme has a specific objective where the goal is for the output of the system, denoted as $y(k)$, to accurately track the reference input denoted as $r(k)$. In addition, for the system to function properly it has to be internally stable, that is, all the signals in the system need to be at least bounded.

Let us recall once again that the motivating factors why we are considering this particular control scheme stem from its successful applications to controlling some of the most complex dynamical systems and for being perfectly suited for the application, that is, classification of time-varying signals which, we believe, can be related to a tracking problem. The internal model control type of control and regulation block diagrams also has history of being used with neural networks both modeling the processes and their inverses. One of these studies resulted in an internal model control scheme that was shown to be effective for a large class of complex nonlinear dynamic systems [13]. The specific modified internal model control scheme, introduced in the previous chapter, has been initially developed using neural networks [24]. The first step of this modification in the context of neural networks, has been obtained by an equivalent input/output transformation of the block diagram as depicted in Fig. 8.2. The more general explanations of the derivations of steps in developing this specific modified internal model control block diagram have been provided in the previous chapter, yet in this chapter they will be only partially replicated and specified toward using neural

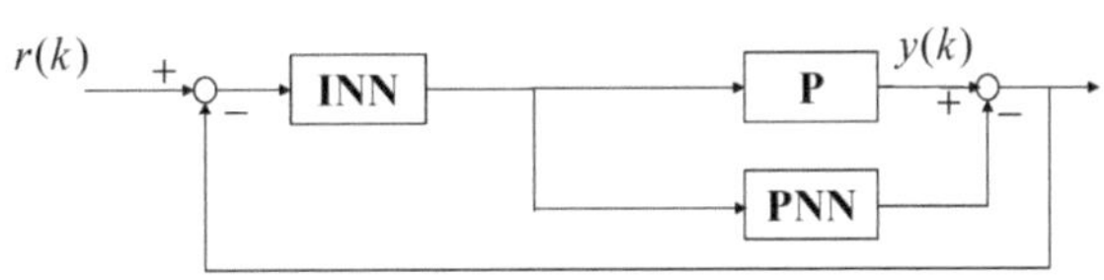

Fig. 8.1 Internal model control block diagram

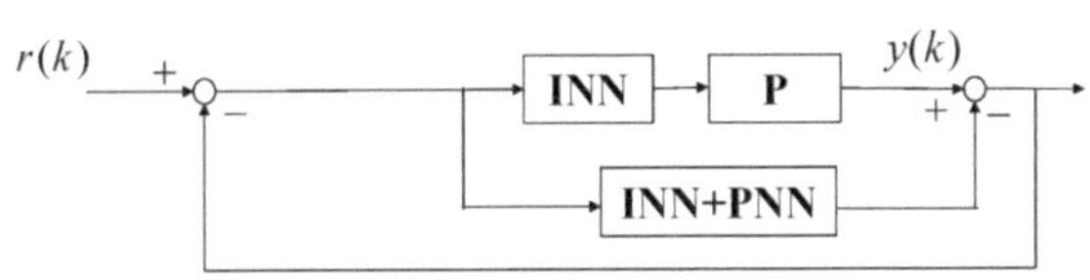

Fig. 8.2 An equivalent input/output modification of the internal model control block diagram

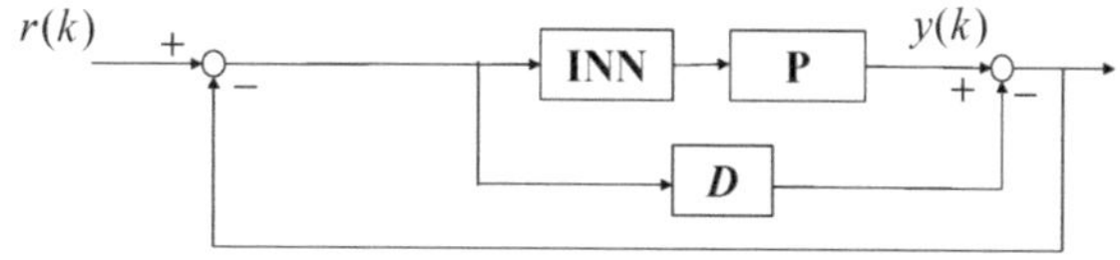

Fig. 8.3 Modified internal model control with neural networks and a delay block

networks not as unknown processes to be controlled but to enhance their tracking and thus classification performance.

Now, the combined block **INN** + **PNN** has been replaced by a delay to accommodate real-time implementation. This simple modification is shown in Fig. 8.3. The inserted delay depends on the estimate of the delay incurred by the process that depends on its dimension and complexity.

Again, in this setup, our goal is not to control or regulate an unknown process but to enhance the recurrent neural network's performance. To achieve that, we first replace the process block with another neural network. This neural network is again denoted as **PNN**, which now stands for any recurrent neural network. This is illustrated in Fig. 8.4.

To enhance tracking capabilities of the recurrent neural network, that is, its classification performance, the combination of neural network blocks **INN** and **PNN** will be dealt with differently than in the previous case when the inverse model was replaced by a controller and the process kept as is. For now, this combination of blocks will be replaced by another neural network block, denoted simply by **NN**. This recurrent neural network within the modified model internal control scheme should be enhanced in tracking capabilities, and thus is expected to perform better in classifying the input time-varying signals. Again, the recurrent neural network itself can be of any general form and may contain a number of layers, and may be trained using any standard procedure. What is furthermore expected is that the block diagram will be well behaved as the whole modified internal model control scheme is input/output bounded.

To provide some technical background, that is, to derive equations for governing the implementation of the modified internal model control concept to enhance the performance of the recurrent neural network, let us assume a simplified (that is, without the filter) block diagram depicted in Fig. 8.5. Let us also assume that the neural network is a single layer neural network governed by a nonlinear time-invariant difference equation

$$y(k+1) = f(y(k), e(k)) \tag{8.1}$$

where

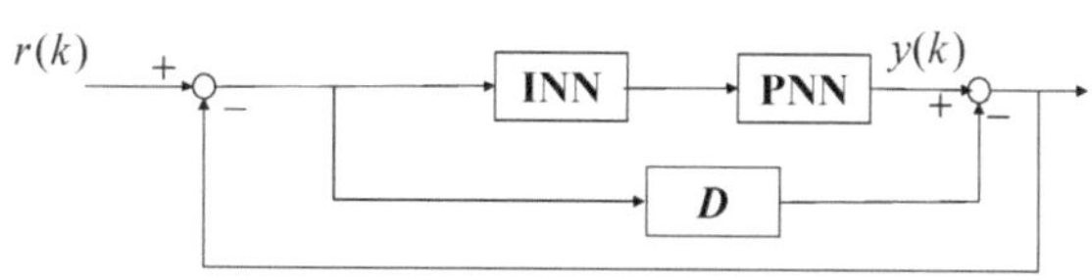

Fig. 8.4 Modified internal model control block diagram with neural networks

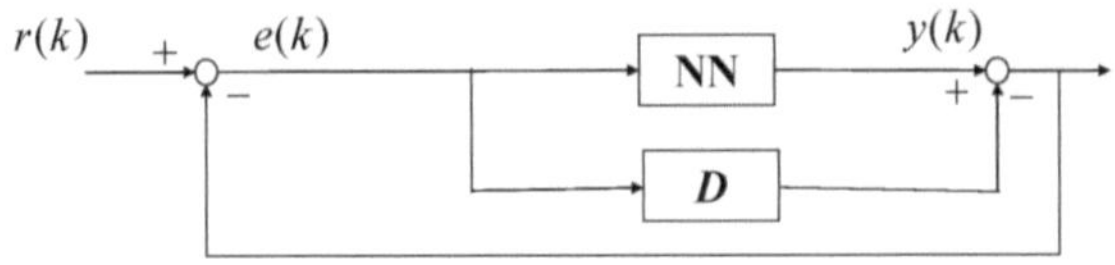

Fig. 8.5 Block diagram for the implementation of a neural network

$$e(k) = r(k) - y(k) + e(k-1) \tag{8.2}$$

so that the overall dynamics simply becomes

$$\begin{aligned} y(k+1) &= f(y(k), e(k)) \\ e(k+1) &= r(k+1) - y(k+1) + e(k) \\ &= e(k) - f(y(k), e(k)) + r(k+1). \end{aligned} \tag{8.3}$$

Difference equation (8.1), representing a nonlinear recurrent neural network model. May stand for a long short-term memory [10, 12, 25], a gated recurrent unit [4], or a light gate recurrent unit neural network model [23], for example.

One important thing to observe is that the equation written in the following form:

$$e(k+1) = e(k) - f(y(k), e(k)) + r(k+1)$$

or the same equation written in an equivalent form

$$e(k) = e(k-1) - f(y(k-1), e(k-1)) + r(k) \tag{8.4}$$

has the integration action incorporated. The block diagram is expected to operate in a feasible way, meaning that the internal signals are expected to be stable. But then we have to be careful given that the integration, while it helps to eliminate or reduce tracking errors, is inherently unstable. Though, the scheme looks appropriate for a small gain theorem analysis, due to the effect of the feedforward branch in tracking, the approach is not perfectly suitable, as is. The most essential condition is that the following sum:

$$\sum_{k=1}^{+\infty} (r(k) - f(y(k-1, e(k-1)) \tag{8.5}$$

converges. In that case, let us assume it converges to e^*. Assuming furthermore, yet without loss of generality, that $e(0) = 0$, we get the following:

$$e(k) \underset{k\to\infty}{\to} e^* \Rightarrow (r(k) - y(k)) \underset{k\to\infty}{\to} 0. \tag{8.6}$$

Here we believe, it has to be emphasized that the asymptotic convergence is not the best way of looking at the tracking performance. Thus, by mitigating the boundary stability property (as known to be a desirable property for the trained recurrent neural networks) to the block diagram and not the neural network itself, the neural network can be designed to be more stable and thus faster to react. So the recurrent neural network within the block diagram could be designed to be faster while the whole block diagram would be taking care of satisfying the "edge of chaos" condition. For the reliability of the overall block operation, it would be desirable to have a simple guarantee, such as an upper bound on discrepancy from the reference signal, that is, as follows:

$$\|y(k) - r(k)\| \leq \epsilon \tag{8.7}$$

for some prescribed value of ϵ. To suppress the instability of the integration action, we would also like to impose the following:

$$e(k+1) - e(k) \underset{k\to\infty}{\to} 0 \tag{8.8}$$

resulting in a bounded error

$$\|e(k)\| \leq \epsilon.$$

Another viable way of treating the problem, is to consider the error equation and perform the following bounding:

$$\begin{aligned}\|e(k+1)\| &\leq \|e(k)\| + \|r(k) - f(y(k), e(k))\| \\ &\leq \|e(k)\| + \delta(k)\|e(k))\| + \beta(k)\|r(k)\| - \gamma(k)\|y(k))\| \\ &\leq (1+\delta(k))\|e(k))\| + \beta(k)\|r(k)\| - \gamma(k)\|y(k))\|\end{aligned} \tag{8.9}$$

where the additional form of bounding is assumed as follows:

$$\begin{aligned}\|r(k) - f(y(k), e(k))\| \leq \;& \delta(k)\|e(k))\| + \beta(k)\|r(k)\| \\ & -\gamma(k)\|y(k))\|\end{aligned} \tag{8.10}$$

for some nonnegative and summable sequences $\{\delta(k)\}_{k=0}^{\infty}$ and $\{\beta(k)\}_{k=0}^{\infty}$, that is,

$$\sum_{k=0}^{\infty}\delta(k) < \infty \text{ and } \sum_{k=0}^{\infty}\beta(k) < \infty. \tag{8.11}$$

If we further assume that the output of the neural network is an output of a sigmoid function, that is, constrained within $[-1, 1]$, then the reference signal needs to be constrained within the same interval, to render the overall problem feasible. Even if the output of the network is not an output of the sigmoidal network we can assume that the reference signal possesses such property. If we take the norm to be the infinity norm, or that simply

$$\|r(k)\| \leq 1, \;\; \forall k \in \mathbb{Z}^+ = \{0, 1, \ldots\} \tag{8.12}$$

then

$$\{\beta(k)\|r(k)\|\}_{k=0}^{\infty} \tag{8.13}$$

would also be a summable sequence. Then, we can conclude by Lemma 5.31 in [1] that $\{\|e(k)\|\}_{k=0}^{\infty}$ is bounded and converges to a nonnegative limit value e^*, implying that the output of the network converges to the reference signal. Furthermore, $\{\gamma(k)\|y(k)\|\}_{k=0}^{\infty}$ is summable, that is,

$$\sum_{k=0}^{\infty} \gamma(k)\|y(k)\| < \infty. \tag{8.14}$$

8.2 Stabilizing the Modified Internal Model Control Scheme with Neural Networks

By introducing delay in the feedforward branch of the modified internal model control block diagram, a realistic and implementable block diagram has been created yet a technical caveat is that the perfect tracking cannot be guaranteed perse. Another important and essential property of this block diagram to be determined is whether it is internally stable, that is, if all signals within the block diagram are guaranteed to be bounded. This feature will be considered when the process is a recurrent neural network that has bounded output. This is a natural assumption for a large class of recurrent neural networks where the outputs are bounded, as a consequence of using sigmoid functions at the output. The usage of a sigmoid function at the output may not be direct like in the case of long short-term memory and gated recurrent unit neural networks, but in such a way that it guarantees the output to be bounded. As it will be shown later, the internal stability analysis would be further enabled if a filter is applied to the error signal. This filter is denoted as **F**, and depicted in the resulting block diagram in Fig. 8.6.

Another advantage of using the filter is to smooth out the signals and remove higher frequency components that may be damaging for the system. An estimate for the delay in the feedforward branch should be based on the recurrent neural network characteristics, such as the number of layers and the type of specific dynamics of the basic network units.

Given that the output of the process in this scenario is the output of the recurrent neural network we assume that they are norm or absolute value bounded by a positive constant, which by no loss of generality will be assumed to be one. This bounding is again a norm like, and for the signals as sequences we assume the ℓ_∞ norm that is

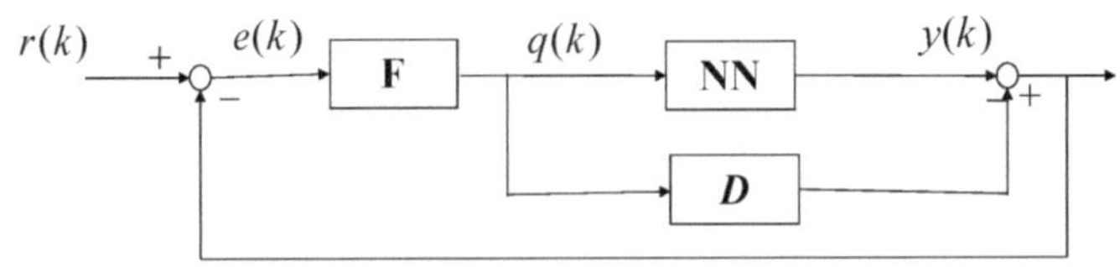

Fig. 8.6 Block diagram for the implementation of a neural network

$$\|x\|_\ell = \sup_{\forall k}\{\|x(k)\|\} \tag{8.15}$$

where $\|\cdot\|$ denotes the standard Euclidean norm of each vector sample at time instance k. Since the design of the filter is completely under our, that is, designer's control, it will be designed to be at least input to state stable. Here, for the simplicity of presentation, we assume that the output of the filter coincides with the state of the filter denoted as q. Notice that the input to the filter is the error signal e. In that case, there exist a $\mathcal{KL}$ function $\beta(\cdot, \cdot)$ and a $\mathcal{K}$ function $\gamma(\cdot)$ such that:

$$\|q(k, q_0, e)\| \le \beta(q_0, k) + \gamma(\|e\|_\ell). \tag{8.16}$$

A simple example of this would be a linear time-invariant system

$$q(k+1) = Aq(k) + Be(k) \tag{8.17}$$

where the system matrix A is Schur stable, that is, it has all of its eigenvalues strictly inside the unit circle. In the case of a stable time-invariant discrete-time linear system, the $mathcal K$-functions in Eq. (8.16) can be computed explicitly as follows [14]:

$$\beta(\rho, k) = c\sigma^k \rho, \tag{8.18}$$

$$\gamma(\rho) = \sum_{j=0}^{\infty} c\sigma^j \|B\| \rho = \frac{c\,\|B\|}{1-\sigma}\rho \tag{8.19}$$

where positive constant c and nonnegative constant σ, with $\sigma \in [0, 1)$, satisfy the following inequality [16]:

$$\|A^k\| \le c\sigma^k, \forall k \ge 0. \tag{8.20}$$

Assuming that the delay in the feedforward branch equals d samples, from the block diagram we get the following relationship:

$$e(k) = r(k) - y(k) + q(k-d). \tag{8.21}$$

Now, we can compute the following:

$$\begin{aligned} \|e\|_\ell &= \|r - y + q^d\|_\ell \\ &\le \|r - y\|_\ell + \|q^d\|_\ell = \|r - y\|_\ell + \|q\|_\ell \end{aligned} \tag{8.22}$$

where q^d represents the delayed sequence q. The inequality (8.22) is obtained since a finite delay equal to the integer number of samples does not change the ℓ_∞ norm of a sequence, resulting in

$$\|q^d\|_\ell = \|q\|_\ell.$$

At this point, we are ready to formulate and essentially finish proving the following theorem.

Theorem 8.1 *If the filter in the modified internal model control block diagram is input to state stable, the reference signal is bounded, and the output of the process is bounded, then a sufficient condition for all signals to be bounded is the following:*

$$\gamma(\|q\|_\ell) \leq \overline{\gamma}\|q\|_\ell, \tag{8.23}$$

where $0 < \overline{\gamma} < 0.5$.

Proof Since $\gamma(\cdot)$ is a strictly increasing function we get the following:

$$\gamma(\|e\|_\ell) \leq \gamma(\|r - y\|_\ell + \|q\|_\ell) \leq \gamma(2c_M + \|q\|_\ell) \tag{8.24}$$

where c_M is a common bound on both reference input and the output of the process. Since both signals are bounded, if the bounds are not the same, which is often the case, then the larger bound is taken as a value for c_M.

If $2c_M \geq \|q\|_\ell$ then

$$\gamma(\|e\|_\ell) \leq \gamma(4c_M) \tag{8.25}$$

and therefore

$$\|q(k, q_0, e)\| \leq \beta(q_0, k) + \gamma(4c_M) \leq \beta(q_0, 0) + \gamma(4c_M). \tag{8.26}$$

If $2c_M < \|q\|_\ell$ then

$$\gamma(\|e\|_\ell) \leq \gamma(2\|q\|_\ell) \tag{8.27}$$

and therefore

$$\|q(k, q_0, e)\| \leq \beta(q_0, k) + \gamma(2\|q\|_\ell). \tag{8.28}$$

Since

$$\|q(k, q_0, e)\| - \gamma(2\|q\|_\ell) \geq (1 - 2\overline{\gamma})\|q(k, q_0, e)\|, \tag{8.29}$$

we obtain

$$\|q(k, q_0, e)\| \leq \frac{1}{1 - 2\overline{\gamma}}\beta(q_0, k) \leq \frac{1}{1 - 2\overline{\gamma}}\beta(q_0, 0). \tag{8.30}$$

Therefore, $q(\cdot)$ sequence is bounded, and thus $e(\cdot)$ sequence is bounded. The output is assumed to be bounded by being the output of a neural network and the reference input is bounded by design. This completes the proof. □

Theorem 8.1 is in the spirit of the small gain theorem [9, 14, 15]. It shows that given the output of the process is bounded by being the output of a recurrent neural network with sigmoid units, only the gain of the loop involving the feedforward branch needs to be small enough. The only issue may be performance wise, that is when the input to the recurrent neural network does not adjust well enough to produce the output satisfactorily close to the reference input. In that case we can put back the controller action into the same branch as the neural network. This is depicted in Fig. 8.7.

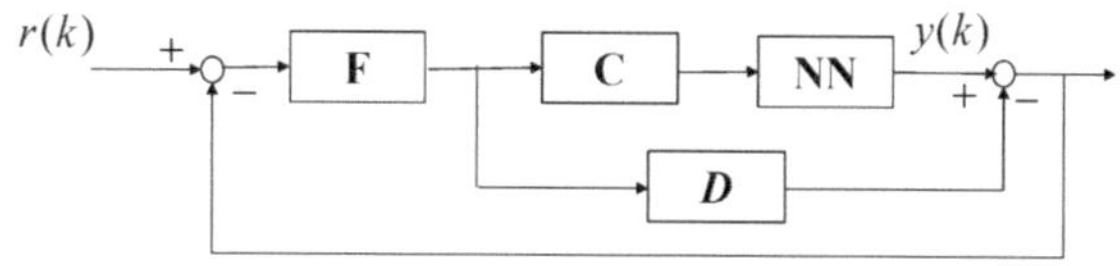

Fig. 8.7 Block diagram for the implementation of a neural network with both filter and controller

As long as the controller itself is input-to-state-to-output stable, all the signals in the modified internal model control scheme for neural networks, should remain bounded. The role of the controller is to produce an input signal that will result in better tracking, and thus better classification of the input sequence.

Example 8.1 For the recurrent neural network block let us assume a basic tangent hyperbolic dynamics in the following form:

$$y(k+1) = \tanh(w_y y(k) + w_r r(k)) \tag{8.31}$$

where $y(\cdot)$ is the scalar output and $r(\cdot)$ is the scalar reference input. For the reference input we choose a superposition of three sinusoidal signals of the following form:

$$r(k) = 0.2\cos(0.5k) + 0.1\sin(0.2k) - 0.15\sin(k) - 0.1 \tag{8.32}$$

where the signal is chosen not to be overly simple, on one hand, and on the other hand not to have maximal value larger than one so that the output of the neural network would be able to track it. It also has a bias so that it imposes another difficulty in terms of tracking.

For the weight parameter values $w_y = 0.9$ and $w_r = 2$, using the modified internal model control block diagram without the filter, we get the output which is plotted together with the reference input in Fig. 8.8.

If we decrease w_y weight value to 0.5, so that the network is faster, the outcome is illustrated in Fig. 8.9.

Further decreasing the value of w_y to 0.1, results in a better tracking, as depicted in Fig. 8.10.

If we include the following linear filter:

$$q(k+1) = 0.2q(k) + 0.2e(k) \tag{8.33}$$

in the modified internal model control block diagram we get even better tracking as depicted in Fig. 8.11.

Therefore, this simple example indicates that the modified internal model control scheme works better for more stable neural networks. This also supports our intuition that the operation at the "edge of chaos" has been mitigated from the recurrent neural network to the modified internal model control scheme.

If we set $w_y = 1$, so we push the neural network dynamics towards instability, then even with the filter we do not get a satisfactory tracking, as shown in Fig. 8.12.

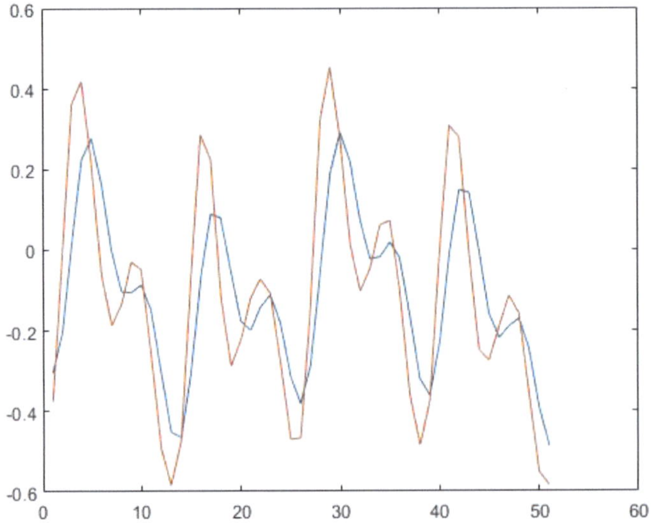

Fig. 8.8 Output (in red) and the reference input (in blue), when $w_y = 0.9$ for the modified internal model control block diagram without the filter

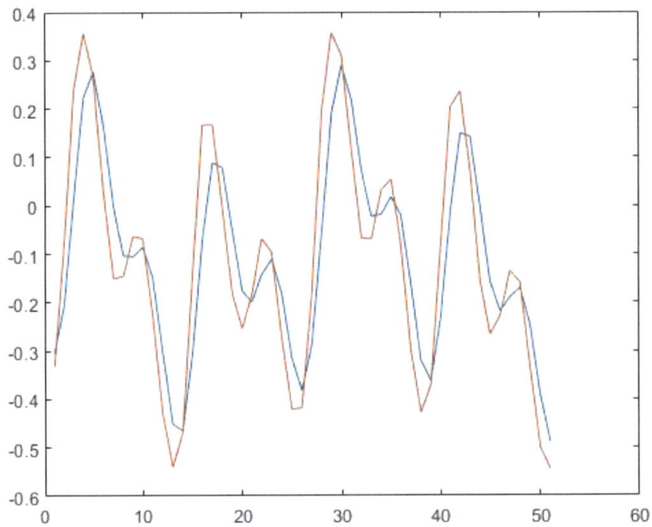

Fig. 8.9 Output (in red) and the reference input (in blue), when $w_y = 0.5$ for the modified internal model control block diagram without the filter

To compensate for destabilizing the recurrent neural network, we can use a slower filter:

$$q(k+1) = 0.5q(k) + 0.2e(k) \tag{8.34}$$

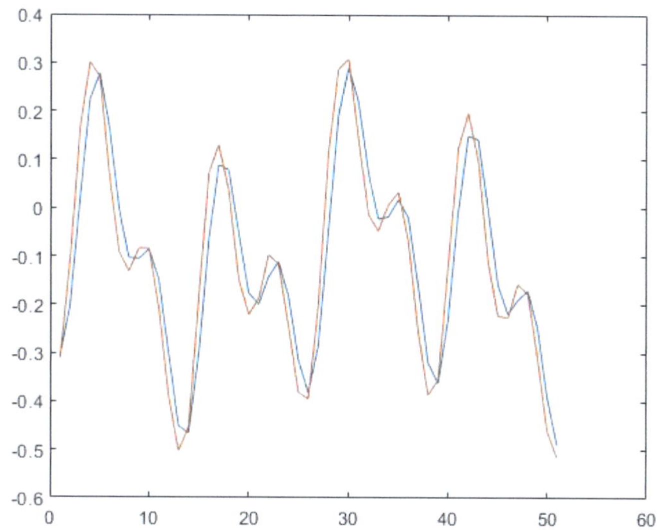

Fig. 8.10 Output (in red) and the reference input (in blue), when $w_y = 0.1$ for the modified internal model control block diagram without the filter

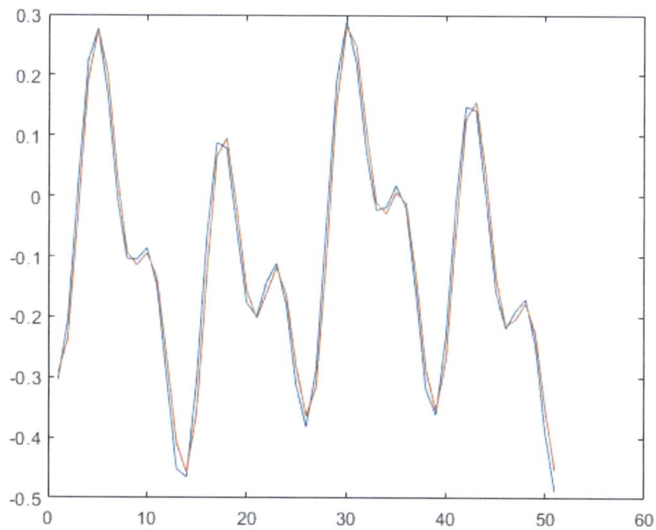

Fig. 8.11 Output (in red) and the reference input (in blue), when $w_y = 0.1$ for the modified internal model control block diagram with the filter

and use the controller to decrease the gain in the main branch. We simply use a reduced gain controller with gain equal to 0.5. So the gain controller serves as an attenuator to reduce the

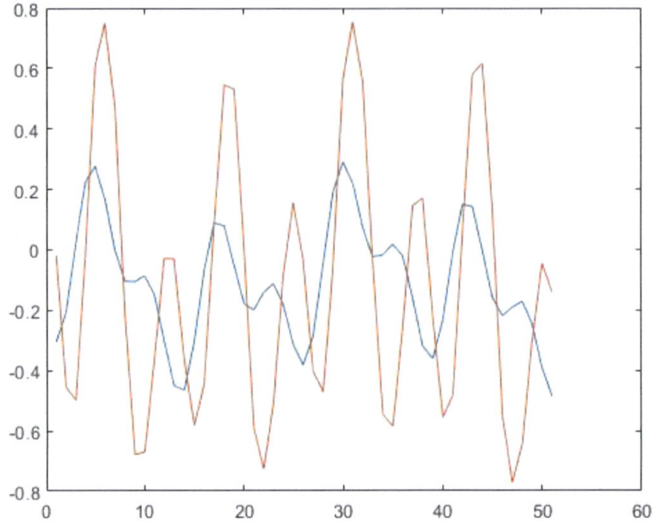

Fig. 8.12 Output (in red) and the reference input (in blue), when $w_y = 1$ for the modified internal model control block diagram with the filter

loop gain, and thus increase stability from the small gain theorem point of view. The result is depicted in Fig. 8.13.

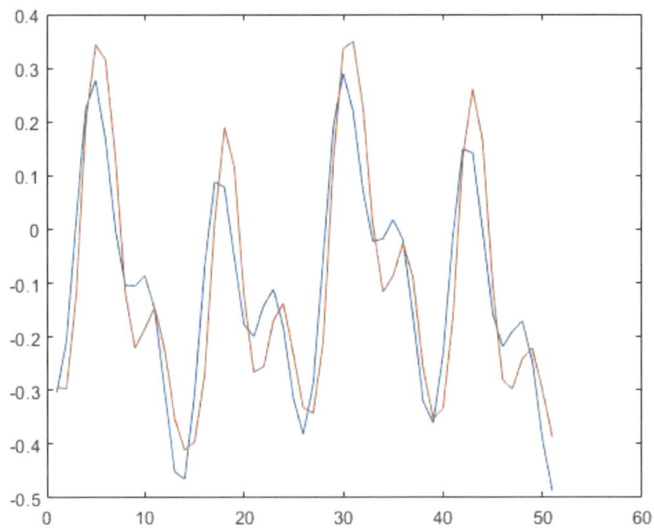

Fig. 8.13 Output (in red) and the reference input (in blue), when $w_y = 1$ for the modified internal model control block diagram with a slower filter and controller gain equal to 0.5

The simulations we have provided, even though the example setups are simple, have demonstrated and indicated various aspects and roles of different blocks in the modified internal model control block diagram with a recurrent neural network. The outcomes of the simulations do support the idea of enabling the use of faster and more stable neural network dynamics while relying on the modified internal model control scheme to enhance the tracking performance. One possible explanation, which we believe in, is that the feedforward branch in the modified internal model control scheme is taking the role related to the operation at the edge of stability, that is, the edge of chaos condition for the well trained recurrent neural networks [2]. If this is true, then there is no need to impose this condition on the recurrent neural network itself. This also may be an explanation why the recurrent neural network performs better, as seen in the provided simulations, within the modified internal model control scheme, when it is faster or more stable. In other words, this would mean that the operation at the edge of chaos is mitigated from the recurrent neural network itself to the modified internal model control scheme. By removing the boundary stable or to operate at the edge of chaos condition, would mean a relaxation in training recurrent neural networks and also reducing the size and the complexity of the training datasets. This may also help with overcoming some well known neural network training issues like overfitting.

The modified internal model control approach to better tracking for recurrent neural networks did not assume any specification on the types of recurrent neural networks. Therefore, it could be any recurrent neural network, with an arbitrary number of layers, and with an arbitrary structure as long as the overall neural network can be represented as a multivariable vector difference equation. This is a consequence of representing recurrent neural networks using difference equation dynamical models. But if we take a look at the neural network as an input-output discrete-time system, it need not be even dynamic, that is, it does not have to include a recurrent component which results in a difference equation. It may just involve the forward information flow which would enable the modified internal model control scheme to be applied to feedforward neural networks like convolutional neural networks [18]. Therefore, technically speaking this approach can include even structures like transformer [26, 27] and informer [28] architectures. But there should be a disclaimer made that if the modified internal model control framework is applied to a feedforward network then the overall structure would impose a feedback information and thus the overall scheme would not be of feedforward nature anymore.

References

1. H. H. Bauschke and P. L. Combettes. *Convex Analysis and Monotone Operator Theory in Hilbert Spaces*. Second Edition, Springer, New York, NY, 2017.
2. N. Bertschinger and T. Natschläger. Real-time computation at the edge of chaos in recurrent neural networks. *Neural Computation*, vol. 16, no. 7, pp. 1413–1436, 2004.
3. F. Bonassi and R. Scattolini. Recurrent Neural Network-based Internal Model Control design for stable nonlinear systems. *European Journal of Control*, vol. 65, 100632, ISSN 0947-3580, 2022.

4. K. Cho, B. V. Merrienboer, C. Gulcehre, D. Bahdanau, F. Bougares, H. Schwenk, and Y. Bengio. Learning Phrase Representations using RNN Encoder-Decoder for Statistical Machine Translation. arXiv:1406.1078, 2014.
5. A. Datta. *Adaptive Internal Model Control*. Springer-Verlag, London, 1998.
6. C. G. Economou, M. Morari and B. O. Palsson. Internal Model Control: Extension to Nonlinear Systems. *Industrial & Engineering Chemistry Process Design and Development*, vol. 25, no. 2, 403–411, 1986.
7. C. E. Garcia and M. Morari. Internal Model Control: A Unifying Review and Some New Results. *Industrial & Engineering Chemistry Process Design and Development*, vol. 21, no. 2, 308–323, 1982.
8. C. E. Garcia and M. Morari. Internal Model Control: Multivariable Control Law Computation and Tuning Guidelines. *Industrial & Engineering Chemistry Process Design and Development*, vol. 24, 484–494, 1985.
9. R. Geiselhart and F. R. Wirth. Relaxed ISS Small-Gain Theorems for Discrete-Time Systems. *SIAM Journal on Control and Optimization*, vol. 54, pp. 423–449, 2016.
10. K. Greff, R. K. Srivastava, J. Koutník, B. R. Steunebrink, and J. Schmidhuber. LSTM: A search space odyssey. *IEEE Transactions on Neural Networks and Learning Systems*, vol. 28, no. 10, pp. 2222–2232, 2017.
11. M. W. Hirsch, S. Smale, and R. L. Devaney. *Differential Equations, Dynamical Systems, and an Introduction to Chaos*. 3rd Edition, Elsevier Inc., Waltham, MA, 2013.
12. S. Hochreiter and J. Schmidhuber. Long Short-Term Memory. *Neural Computation*, vol. 9, no. 8, pp. 1735–1780, 1997.
13. K. J. Hunt and D. Sbarbaro. Neural networks for nonlinear internal model control. *IEE Proceedings-D*, vol. 138, no. 5, pp. 431–438, 1991.
14. Z. -P. Jiang and Y. Wang. Input-to-state stability for discrete-time nonlinear systems. *Automatica*, vol. 37, pp. 857–869, 2001.
15. Z. P. Jiang, Y. Lin, and Y. Wang. Nonlinear small-gain theorems for discrete-time feedback systems and applications. *Automatica*, vol. 40, pp. 2129–2136, 2004.
16. J. P. LaSalle. *The Stability and Control of Discrete Processes*. Springer-Verlag, New York, NY, 1986,
17. G. A. Leonov. *Strange Attractors and Classical Stability Theory*. St. Petersburg University Press, St. Petersburg, Russia, 2008. LeCun Y, Bengio Y, Hinton G. Deep learning. https://doi.org/10.1038/nature14539. PMID: 26017442.
18. Y. LeCun, Y. Bengio, and G. Hinton. Deep learning. *Nature*, 521(7553):436-44, 2015.
19. M. R. Mataušek and D. M. Stipanović. A New Approach to Nonlinear Control System Design. *Journal of Automatic Control*, vol. 5, pp. 31–42, 1994.
20. M. R. Mataušek and D. M. Stipanović. Modified Nonlinear Internal Model Control. *Control and Intelligent Systems*, renamed to *Mechatronic Systems and Control*, vol. 26, pp. 57–63, 1998.
21. M. Morari. Internal Model Control - Theory and Applications. *IFAC Proceedings Volumes*, vol. 16, pp. 1–18, 1983.
22. M. Morari and E. Zafiriou. *Robust Process Control*. Prentice Hall, Englewood Cliffs, NJ, 1989.
23. M. Ravanelli, P. Brakel, M. Omologo, and Y. Bengio. Light gated recurrent units for speech recognition. *IEEE Transactions on Emerging Topics in Computational Intelligence*, vol. 2, no. 2, pp. 92–102, 2018.
24. D. M. Stipanović. *Controlling Nonlinear Systems using Neural Networks*. Diploma Thesis, School of Electrical Engineering, University of Belgrade, Belgrade, Yugoslavia, 1994.
25. D. M. Stipanović, B. Murmann, M. Causo, A. Lekić, V. R. Royo, C. J. Tomlin, E. Beigne, S. Thuries, M. Zarudniev, and S. Lesecq. Some local stability properties of an autonomous long short-term memory neural network model. *Proceedings of the 2018 IEEE International Symposium on Circuits and Systems*, 2018.

26. R. E. Turner. An Introduction to Transformers. arXiv:2304.10557, 2024.
27. A. Vaswani, N. Shazeer, N. Parmar, J. Uszkoreit, L. Jones, A. N. Gomez, Ł Kaiser, and I. Polosukhin. Attention is All You Need. *Proceedings of the 31st International Conference on Neural Information Processing Systems*, pp. 6000-6010, 2017. Longer and revised version available as arXiv:1706.03762.
28. H. Zhou, S. Zhang, J. Peng, S. Zhang, J. Li, H. Xiong, and W. Zhang. Informer: Beyond Efficient Transformer for Long Sequence Time-Series Forecasting. *Proceedings of the Thirty-Fifth AAAI Conference on Artificial Intelligence (AAAI-21)*, pp. 11106–11115, 2021.

Concluding Thoughts

What one wants to accomplish and what one has accomplished may be two different things, which may not necessarily be a bad thing so let it be mentioned and left at that. In a perfect world, as a technical monograph this work should be in order, systemized, self sufficient as much as the restrictions allow, and neatly presented. But beside the subjective obstacle that this monograph is an unavoidable self reflection of the author, there are also some objective obstacles. First given the technical tools and their capabilities, it should not be an exaggeration to say that only "bits and pieces" of general problems related to dynamical systems can be solved and fully understood. For this reason, a restrictive set of feasible problems and related technical tools and ideas have been presented often as a combination and somewhat a balance between systematic and chronological. The motivation for these problems and tools is also a combination between their importance and relevance on one hand, and author's professional expertise, experiences and research interests, on the other. There is a hope that this imperfect work or parts of it will still be inspiring and helpful.

What can one say when the things are done. Not to look back is certainly an option. The other is to comment on what has been done, and based on that to offer suggestions for the future research endeavors. I will choose a hybrid decision, so again some kind of a compromise or in a light of this book some kind of a convex combination. I will not comment on what has been provided in technical terms but offer some suggestions based on what has been presented on how to move forward, in a general aspect. I think it is important we do not rely so much on the brute force approach based on powerful numerical computations, that often do not provide much insights into and explanations for the behaviors of dynamical systems. We need to take a step back and allow ourselves time to think. Yes, this is an option without guarantees but isn't that the way all great ideas were born. Here born is intentionally used rather than created because the great ideas, and ideas in general, do come after hard work and dedication, but in mostly unexplainable, unexpected and above all uncontrolled moments. And this book is about control so yet another contradiction is encountered. This

D. Stipanović, *Difference Equations and Machine Learning*, Synthesis Lectures on Mathematics & Statistics, https://doi.org/10.1007/978-3-032-00910-4

is not to be ignored as we often and most commonly do, but rather has to be nurtured and respected. The expected outcomes even though unpredictable, which is yet another contradiction, would not only provide explanations for behaviors of dynamical systems but also lead to less computationally intensive and more meaningful ways of controlling and regulating them. In other words, it would be of preference to seek more solutions in explicit forms with well defined guarantees. Developing new tools and theories would be unfair to suggest yet a need for them becomes more evident so at least we should create environments to give them a fair chance of being born and nurtured. In general, compromising between and utilizing general theories and particular scenarios seems to be an inevitable way of dealing with dynamical systems. More than a few contradictions are mentioned, yet one should not avoid them but embrace that contradictions are a natural phenomenon arising from many problem formulations. Chaos being desirable may be taken as one example. This monograph as another.

Glossary

Artificial intelligence Artificial intelligence is based on algorithms performing tasks that typically require human intelligence. These tasks include learning, reasoning, problem-solving, perception, and decision-making. The algorithms are programmed to learn from the available datasets, adapt to new situations, and provide meaningful decisions.

Difference equation A difference equation is an equation that defines the relationship between consecutive terms in a sequence. It is the discrete analog of a differential equation, meaning it deals with changes in a variable at discrete time steps rather than continuously.

Edge of chaos Boundary of a region where the corresponding dynamical systems manifest chaotic behaviors.

Gated recurrent unit neural network A gated recurrent unit neural network is a type of recurrent neural network. It is obtained by a simplification of the long short-term memory neural network.

Homotopy The word "homotopy" comes from Ancient Greek, combining homos, meaning "same, similar" and topos, meaning "place."

Internal model control Internal model control is a model-based control technique that uses both process model and its inverse to enable the process to track the reference signals while attenuating the disturbances.

Long short-term memory neural network A long short-term memory neural network is a type of recurrent neural network. These neural networks are predominantly used to learn, process, and classify sequential and time-varying data because they can learn long-term dependencies between time steps of data.

D. Stipanović, *Difference Equations and Machine Learning*, Synthesis Lectures on Mathematics & Statistics, https://doi.org/10.1007/978-3-032-00910-4

Lyapunov theory A theory that provides theoretical guaranties for solutions of differential and difference equations. The guarantees are formulated such that they do not need these equations to be explicitly solved.

Machine learning Machine learning enables using digital computers to learn from the provided datasets. It is based on algorithms that can analyze data, identify patterns, and make predictions or decisions based on those patterns and datasets.

Recurrent neural network A recurrent neural network utilizes recurrent connections for the individual neurons.

Index

D. Stipanović, *Difference Equations and Machine Learning*, Synthesis Lectures on Mathematics & Statistics, https://doi.org/10.1007/978-3-032-00910-4

MIX
Papier aus verantwortungsvollen Quellen
Paper from responsible sources
FSC® C105338

If you have any concerns about our products,
you can contact us on
ProductSafety@springernature.com

In case Publisher is established outside the EU,
the EU authorized representative is:
Springer Nature Customer Service Center GmbH
Europaplatz 3, 69115 Heidelberg, Germany

Printed by Libri Plureos GmbH
in Hamburg, Germany